Disarming Hitler's V-Weapons

Disarming Hitler's V-Weapons

Bomb disposal: the V1 and V2 rockets

Chris Ransted

Pen & Sword
MILITARY

First published in Great Britain in 2013 by
Pen & Sword Military
an imprint of
Pen & Sword Books Ltd
47 Church Street
Barnsley
South Yorkshire
S70 2AS

Copyright © Chris Ransted 2013

ISBN 978 1 78159 386 8

The right of Chris Ransted to be identified as the Author of this Work has been asserted by him in accordance with the Copyright, Designs and Patents Act 1988.

A CIP catalogue record for this book is available from the British Library

All rights reserved. No part of this book may be reproduced or transmitted in any form or by any means, electronic or mechanical including photocopying, recording or by any information storage and retrieval system, without permission from the Publisher in writing.

Typeset in Ehrhardt by
Mac Style, Driffield, East Yorkshire
Printed and bound in the UK by CPI Group (UK) Ltd, Croydon, CR0 4YY

Pen & Sword Books Ltd incorporates the imprints of Pen & Sword Archaeology, Atlas, Aviation, Battleground, Discovery, Family History, History, Maritime, Military, Naval, Politics, Railways, Select, Social History, Transport, True Crime, and Claymore Press, Frontline Books, Leo Cooper, Praetorian Press, Remember When, Seaforth Publishing and Wharncliffe.

For a complete list of Pen & Sword titles please contact
PEN & SWORD BOOKS LIMITED
47 Church Street, Barnsley, South Yorkshire, S70 2AS, England
E-mail: enquiries@pen-and-sword.co.uk
Website: www.pen-and-sword.co.uk

Contents

Acknowledgements		vi
Introduction		1
Chapter 1	June 1944: The First V1s	13
Chapter 2	More V1 'Duds'	42
Chapter 3	Unexploded V2s in England	67
Chapter 4	Europe	87
Chapter 5	Tools of the Trade	134
Chapter 6	Post-War Discoveries	167
Chapter 7	Fact or Fiction?	191
Epilogue		201
Notes and Sources		208
Appendix 1	George Medal Awards (V1 incidents)	220
Appendix 2	V1 Disposal Instructions	222
Appendix 3	V1 Fuze Technical Information	244
Appendix 4	V2 Disposal Instructions and Nose Fuze Technical Informatiown	248
Index of People, Places and Units		251

Acknowledgements

Individuals

Roy and Doreen Ashmeade
John D. Bartleson Jr
Noel Cashford
Bert Clinch
Kim Freeman
Robin Gaymer
John Glascock
Wolfgang Gückelhorn
Hermann Hinsenveld
Rob Hoole
John Hudson
Richard Hudson
Tom Ledger
Volker Lessmann
Raymond Maries
Antoon Meijers
Lionel Meynell
Karen Nicholls
Geoff Nutkins
Detlev Paul
John Pridige
Winston Ramsey
David Ransted
Dennis Reece
Martin Russell
Ray Scott
Eric Sivil
Jochen Tarrach
Hans Joachim Ulmer
Steve Venus
Lieutenant Colonel Eric Wakeling (Ret'd)

Organizations/Museums

Imperial War Museum
Minewarfare & Clearance Diving Officers' Association
The National Archives
The Netherlands EOD Foundation Historical Collection
Norfolk and Suffolk Aviation Museum
RAF Bomb Disposal Association Robertsbridge Aviation Society
Robertsbridge Aviation Society
Royal Engineers Bomb Disposal Association
Wings Museum

Introduction

Fear of the unknown is something familiar to all of us. To a bomb disposal officer in Britain during the Second World War this fear was very real. The consequences of trying to investigate a bomb the construction of which was unknown could result in disaster.

Put yourself in their shoes and imagine for a moment that you have been called to the scene of an unexploded bomb – perhaps an unexploded V1 flying bomb or V2 rocket. Your job is to make it safe. You may never have seen one up close before and know little of the details of its workings. Just being in close proximity to a bomb is frightening enough, but what about those unknown factors?

Even if you have read all the technical circulars provided by the backroom boffins put together from the experiences of others, this bomb might be different. It could be an updated version with a slightly different design, or with booby-traps designed specifically to kill you and keep the weapon's technical details safe. It might be fitted with a simple mechanical switch, like the one that turns the light on in your fridge as you open the door. This could be activated as you remove a component or undo a bolt. Possibly there is a trembler switch sensitive to the slightest movement or vibration. Alternatively it might have a photo-electric cell waiting to create an electric current to the detonator as soon as you begin to get inside the workings and expose them to the light of day. There could be a timer ticking away the seconds to detonation while you deliberate over your next move. Or there might be a chemical as opposed to a clockwork time delay, and while you are scratching your head wondering, the acid from a glass ampoule smashed on impact could be gradually eating through the only thing holding back a spring-loaded booby-trap. Even if there are none of these things, panic might take control of you. Will your mind go blank or will you remember all the procedures laid down for disarming the bomb? Are there such procedures, or are you the one tasked with writing them through your own 'trial and error' explorations? (You probably wouldn't make a lot of errors – one would be enough, and that would be written in to the procedures for the next person.)

How about the stories you've been hearing, like the one about the Navy bomb disposal officers working on acoustic sea mines, worrying that it could detonate if too much noise was made with their spanners, remembering to remove the metal buttons on their trouser flies and all other metal about their person in case it altered a mine's finely tuned magnetic balance and set it off. (One naval officer took all these precautions only to find someone had left a bicycle leaning against the parachute mine he was tasked with disarming, while another found his mine lying across a railway track, having welded itself to the live rail.)

Having crashed to earth the bomb could be damaged, a tangled mess of metal, explosive, and wires – the insulation on these possibly damaged with an unnoticed piece chafed through, its bare spot just a fraction of an inch away from making contact with another metal surface and thereby short-circuiting and detonating the bomb. How steady are your hands? Could you wrap insulating tape around a wire without disturbing the one next to it?

Maybe there is an air raid going on around you, or you are being shot at while you try to work. What if the bomb you are dealing with has fallen in the middle of a minefield?

While working on your bomb you might hear the sudden explosion of another UXB nearby, its timer running to its conclusion and reminding you that your bomb could blow at any second. Perhaps you knew your friend had been working on that other bomb that you just heard explode.

It might be that you are trying to work 'blind' – by feel alone, up to your waist in mud, freezing water, or sewage. Your hands are wet. What if the water causes a short circuit? Maybe both you and the bomb are completely immersed in water at the bottom of a dock. You are hanging upside down in the darkness in a cumbersome diving suit, with the current pulling at you. All you can hear is the sound of your own breathing and your heart beating. The water is icy cold and your hands numb. If you drop any of your tools in the swirling silt you would probably never find them again.

The environment you are forced to work in could be quite toxic, as when Lieutenant Hugh Cronyn had to deal with a 500kg UXB that had come to rest in the fume-filled fuel tank of the SS *Chesapeake* in the Bristol Channel.[1] To work on that bomb he had to wear a wooden box over his head and have fresh air pumped to him down a length of hose. Or there was the bomb in the gasometer in London's Kings Cross, where Captain A. G. Polson had to climb down the long ladder to it wearing a deep sea diver's suit.[2]

And what of the other external pressures? Suppose you have had hardly any sleep for days – spent the past few weeks constantly working on unexploded bombs – your nerves strained to breaking point. Perhaps there are problems at home, too. Could you still focus?

There might be hundreds of people waiting on you – praying you are successful and their homes will be safe. The factories are waiting to resume their vital war work, communications on hold, railway lines shut down, roads and telephone exchanges – the boffins too, are waiting to get their hands on components that might unlock the secrets and find a way of defeating this weapon. You may be the person who will find the necessary clue and thereby prevent death and destruction in the future. It all depends on you! How well do you think you could handle the pressure?

Lieutenant Eugene Haderlie was a US Navy officer attached to the Royal Navy, who spent a good part of the war in a diving suit, disarming sea mines at the bottom of the English Channel. He summed up his thoughts on this 'challenge' quite succinctly: 'The more you know, hopefully the better off you'd be. The difference between the kind of work that I did and the work that a GI soldier would do was simply that the soldier was up against another man who was probably about his equal in every way. But the mines we were dealing with had sophisticated mechanisms that were designed by some of the best engineers and physicists in the entire world. And we felt that we were at a disadvantage, let's say.' The same could easily be said by those who had to work on V-weapons.[3]

All the dangerous scenarios described above were real ones, faced by real people, by bomb disposal personnel in the course of their duties. Contrary to popular belief you did not have to volunteer to be in bomb disposal. The men doing this work were, before the war, ordinary men doing ordinary jobs. They found themselves presented with this challenge. Hopefully this book will give some insight into the work they performed.

The V2 rocket was a precursor of modern space travel technology. It can be traced back to the early 1930s when rocket enthusiast clubs sprang up around Germany. Hitler knew rockets had been overlooked by the Versailles Treaty at the end of the First World War, so their development could be legitimately exploited. As the storm clouds gathered over Europe he was well aware of the potential for rockets as machinery of war.

Aerial photos and spies in continental Europe had discovered the existence of the Nazi 'Vengeance' weapons before any had arrived on British shores. Unfortunately little has been written about bomb disposal personnel's role in unlocking the secrets of these weapons. Yet their endeavours should certainly not be overlooked; they were not just risking their lives to disarm an individual weapon but were always trying to find out how they worked, to find ways of defeating them. Their ultimate mission, we can now appreciate, was to try to stop the forerunner of what would later be known as the 'intercontinental ballistic missile'.

Wild rumours often circulated about 'new' secret weapons that Hitler was about to unleash – 'death rays' and so on. The press didn't help matters.

For example, the *Evening Standard* of 17 December 1943 mentioned a 'suffocation bomb' that could destroy oxygen over a large area. To acquire real first-hand knowledge and a strategy for dealing with any new threat was of paramount importance and could only be good for public morale.

These V-weapons were cutting edge and, certainly in the case of the V2s, complex technology. The parts required precise engineering tolerances. Components were used from a number of different manufacturers and often had to be 'adjusted' as they were assembled (the V2 actually had over 20,000 individual parts).[4] The powers

A number of newspapers covered this story of the V2 being a 'glacial bomb'. This particular extract is from the *Western Mail*, 31 August 1944. It would be just over a week later when the reality of the V2 began to be understood by the British public. (*Author collection*)

This photo gives a good idea of the destructive power of a V1. Early in the campaign this one laid waste a typical London street. Censorship at the time was such that precise locations were not given in the press to prevent the Germans from gauging the weapon's accuracy. (*Author collection*)

that be in Germany must have realized this would pose a problem, besides which Hitler tried to rush these wonder weapons into service, with not quite all the bugs ironed out. Because slave labour was involved in their manufacture, sabotage also occurred – welds that were hidden might only be partially completed, or there might be poor soldering on electrical connections. It has also been recorded that slave workers urinated and defecated into the fuel tanks, or put pieces of paper into the fuel tanks and turbo pumps, causing misfires. Instances of this last act became so common that a report was drafted in December 1944 stating that 'papers' would be examined to try to identify the culprits. One member of a V2 launch crew recalled that 2 out of 10 V2s had technical problems literally due to 'spanners in the works', or sometimes holes in the outer skinning, perhaps after being shot at.[5] To put a stop to sabotage at the factories the Germans made the labourers put named slips of paper with their work so it was their responsibility if it failed. The penalty for sabotage was death. Cranes in the V2 factory at Nordhausen were used to hang suspected saboteurs in full view of the workforce (and the scientist who developed the rockets) – the victims left hanging throughout the working day. A permanent gallows was also erected in the roll call yard. It is said that over 200 workers were publicly hanged for suspected sabotage,[6] and thousands of others at Nordhausen were worked, starved or beaten to death.

The threat of death did not stop sabotage – some very brave acts were undertaken by unnamed forced workers in the fight against the Nazis. Captain F. Ashe Lincoln, an expert in rendering safe parachute mines and other naval ordnance, recalled an incident involving one of his colleagues.[7] A magnetic mine was found unexploded. It looked perfectly normal but when the officer began to disarm it he found that the internal workings had been sabotaged. Inside the mine's casing a slave labourer had written, 'We are with you' and drawn the Star of David. This was apparently reported to Winston Churchill, who ordered that the incident should not be made public. This of course was not only to avoid reprisals on slave labour, but also because such help from those working in German munitions factories could save the Allies many lives.

It should come as no surprise bearing in mind the circumstances of manufacture that some V-weapons did fail. Some details of the weapon's failings were obtained from German prisoners of war captured after D-Day. One individual who was involved in the testing of V2s described how after a period of theoretical training a group of battery personnel were sent to Heidelager for their first 'live' firing. Seven rockets were to be launched but three failed, including one where the tail portion – including the burner – blew off in the air about seven seconds after launching. A fourth rocket had technical problems but was launched anyway. Another battery had even more failures. Having unsuccessfully attempted to launch a dozen rockets they were forced to move due to the approach of Russian troops.

The prisoners of war provided a list of common reasons for V2 failures, which included:

1. Icing up of pressure-reducing valve and relays. (These were heated by hot air before launching.)
2. Low temperature affecting electrical resistance valves.
3. Jamming of electric servo motors that controlled the air tabs on the fins.
4. Extremes in temperature affecting the viscosity of oil in hydraulic servos.
5. Irregular supply of the peroxide and permanganate.
6. Burner explosions due to vaporized fuel within the rocket.
7. Fuel tank explosions.

Any information regarding V-weapon failures obtained by the Allies would be exploited to the full. Any detail, however small, that might help defeat this weapon was explored. There was an instance where a bomb disposal officer, about to steam out the explosives from one of the first unexploded V1s, noticed some tiny fragments of rock embedded in the explosive filling. He thought they might be of significance. They were subsequently sent off to a geological expert, Doctor James Phemister from the Government Chemist Department, who had them sectioned and analysed.[8] They were hoping to pinpoint an area in Europe where the stones originated, perhaps indicating where the V1 had had its warhead filled. From such information it might have then been possible to have agents or a photographic reconnaissance aircraft search a localized area,

This photo shows evidence of V1 failures. The launch ramp is in the top right corner. Notice the line of skid marks across adjoining fields where a number of V1s crashed soon after being launched from this site in the Pas de Calais area. (*After the Battle*)

Evidence of some V-weapon failures can still be seen today. This crater was created by a V2 that crashed back to earth near its launch site at Rossbach in Germany. (*www.V2rocket.com*)

leading to further intelligence or even a target for bombing. As it turned out a specific area could not be determined from the samples, but it shows the degree of detective work that went on by the 'backroom boys'.

The authorities were open to any suggestions in the fight against V1s, or Pilotless Aircraft (PAC) as they were known in the early days. Many people came up with ideas, including one from a nine-year-old boy who suggested the use of metal nets, held in the sky by pilotless helicopters.[9]

Brigadier Bateman, Director of Bomb Disposal, was sent a note of one suggestion just two days after the first unexploded V1 was found. It described a telephone call received by the Home Office Police duty room from a member of the public whose idea was that a fast-flying aircraft should chase the flying bomb. The pilot would manoeuvre his machine so that his passenger (a bomb disposal expert) could climb on to the flying bomb and remove the fuze or otherwise render it harmless, before descending by parachute. Before jumping off, it was suggested that he should radio anything he had discovered to ground headquarters. The Home Office added their own sarcastic comment to Bateman, to the effect that they were sure some of his bomb disposal officers would be putting on their riding boots and spurs forthwith![10] Other ideas were taken more seriously, such as the possibility of somehow scattering paper in the missile's path in an attempt to choke up the intake to the jet engine and cause it to stall.[11]

It would be natural to assume that the Nazis would want to keep the knowledge of how the V-weapons worked a closely guarded secret.

Incorporating a booby-trap would be an obvious safety measure should the weapon fail on impact. That would not only hopefully destroy any physical evidence, but also kill the bomb disposal officer, removing some expertise from the pool of British resources.

This idea was not novel. A deliberate attempt to kill British bomb disposal experts with a new booby-trap occurred in August 1940, when a parachute mine was found unexploded at Bere Farm, Boarhunt. It actually had three booby-traps, one of which evidently exploded as the mine hit the ground, blowing open the casing without detonating the main charge. Commander Geoffrey Thistleton-Smith was called to the scene and found there was still work to be done – the end of the mine with the detonator and primer remained intact. The next day Commander John Ouvry removed these and recovered all the exploded fragments he could find. On further investigation it was found that technically, it was not a mine at all. There was no clock or magnetic unit, which meant it could not have been dropped with the aim of blowing up shipping. All that was fitted were elaborate booby-traps designed to kill the bomb disposal officer and anyone else close by.[12] The fact that it was dropped so close to the home base of the Naval mine-disposal experts, HMS *Vernon* at Portsmouth, was probably no coincidence.

In fact another sea mine was reported on the same day, dropped some ten miles from the sea, a mile and a half from Piddlehinton. The mine had initially landed at the top of a sloping grassy meadow, but rolled some distance with its parachute still attached. Commander Thistleton-Smith, along with an officer by the name of Anderson and an Admiralty boffin, Leonard Walden, inspected the mine and found the normal mechanism for releasing the primer and the plate covering the detonator were in place, but there was nowhere for a clock or bomb fuze. The mine was rolled over (it had rolled down a hillside already, so this wasn't thought to be too risky) and photographed from all angles. Anderson removed the detonator and primer, assuming any booby-traps would be hidden in a less obvious place, and a trepanner used to cut circular holes was fitted to the mine's casing. This tool developed by the National Physical Laboratory was made of non-magnetic materials and driven by compressed air. After it had cut most of a 4-inch circular hole, Leonard Walden cut through the remainder by hand with a hacksaw blade, being careful not to let the blade go too deep into the mine's casing. This exposed the end of the battery power source and the leads from it. The wires were cut and insulated, thereby rendering the electrically operated booby-trap safe. However, it was believed that a mechanically operated one was still in place at the rear of the mine. This area would be difficult to drill through with the trepanner, as there were strengthening ribs in the way.

The men at this point had already spent six days working on the mine and it was decided to use plastic explosive to open up the rear door. Chief Petty Officer Thorns who was in attendance with a small working party was given this 'enjoyable' task. Having lit the fuze and joined the other three men in a slit trench dug in an adjoining field, they waited for the bang. After the plastic explosives went off, the men waited a few seconds more before approaching the mine, which was about 250 yards away. Thistleton-Smith was in front and only about 50 yards from the mine when it suddenly exploded.

Pieces of mine and clumps of earth showered down around the men, who were now flat on the ground. The heavy battery landed only a yard in front of Thistleton-Smith, and the weighty parachute shackle fell on their lorry, 100 yards away. It appeared that the booby-trap had worked but they were not sure of the reason for the delay. All the pieces were gathered up and taken to HMS *Vernon* for analysis.[13] (Thistleton-Smith, Walden and Able Seaman William Comfort were awarded George Medals for their work on these mines.)[14]

It may be the Germans were over confident with their V-weapons and truly believed that there would be no unexploded examples, or perhaps they were just careless, as history has shown that there were actually no booby-traps hidden inside the weapons. They obviously did have some concerns at the time of launching, as the V2's fuzes were not to be armed until 40 seconds after launch.[15] This was supposed to be some protection for the launch crews if anything went wrong. (Though if you have ever seen the footage of a V2 toppling over at launch, with the fuel spilling out and then detonating, you will know that the warhead not exploding would probably be of little consolation to the launch crew!) It was also possible for the crew to shut down a V2 engine via a radio signal,[16] if the engine was either not producing full thrust just before launch, or if it was already airborne but off course. To cut the engine of a missile in the air required some good guesswork as to where the stalled rocket was likely to fall. The V1 also had some safety devices built in, including a mechanism for arming one fuze only after it had travelled about 40 miles and another 6–8 minutes after launch.

That V-weapons did not contain booby-traps takes nothing away from the bomb disposal men's efforts. They could not know, and such devices were fitted as a matter of course in many 'unexploded' bombs dropped from aircraft. Besides which the men were still dealing with bombs in a highly dangerous state – they had crashed to earth, armed, with fuzes that were possibly damaged or in a delicate condition. They could explode at the slightest tap. And of course the warheads that surrounded these fuzes contained a whole lot of explosive (nearly a ton of the stuff). The fact is, many bomb disposal men, real people who had families, friends and colleagues, were killed while working with V-weapons.

10 Disarming Hitler's V-Weapons

This photo dated 20 June 1944 was used with an intelligence report on V1s. Judging by the bombs lined up against the wall on the left, it was taken after the V1 had been retrieved by a BD unit. The only unexploded V1 known to have been recovered by that date was badly smashed up, so this one might be from an exploded example. Sometimes the wreckage could look misleading. A V1 that fell on 23 June 1944 at Salehurst in Sussex was described in a report of the time as 'exploded but nearly perfect specimen apart from warhead'. (Photo: The National Archives: ref.HO 199/465) (*Author collection*)

The policy for dealing with a first-of-its-kind unexploded bomb was to make copious notes at each stage of the 'disposal' operation. Then if the bomb went off and the BD officer was killed, at least there would be a record left to assist the next officer tasked with trying to disarm a similar device. Some of these records have fortunately survived and, coupled with other sources such as diaries and recollections of those involved, form the basis of this book. It should be noted that the author was born twenty years after the war so has no first-hand knowledge – I can only rely on the testimony of others. Some errors may come to light. Even records written at the time may have contradictory details, not to mention crossings out and corrections.

In this book I have tried to include as many names as possible of those involved, as well as photographs of them, to remind readers that these were real people and to bring this history to life a little. I may appear at times to go off on a tangent, detailing bomb disposal work on other types of ordnance, but

Introduction 11

A V2 that fell back to earth shortly after launch from railway sidings at Karlshagen, near Peenemünde. (*German Federal Archive, RH8II Bild-B1976-44*)

I feel it is important to see the BD men's work on V-weapons in context. The 'rockets' were only one element of these men's day-to-day responsibilities, and the work they performed on other types of German weapons gave them vital experience and improved techniques that could be applied across the board to *all* of Hitler's arsenal.

Lieutenant Bassett was killed when a fuze from an unexploded V1 detonated while he was investigating the crash site. He is buried in his home town of Westerham, Kent. (*Author collection*)

I have of course included a lot of technical details of the work involved, and although some of this could literally be described as 'rocket science', I have tried to provide as much information as possible in the text without turning the book into some sort of technical manual. However, in the interests of completeness I have included in the Appendixes the official 'instructions' to BD Officers for disarming both the V1 and V2.

In my previous book, *Bomb Disposal and the British Casualties of WW2*, I detailed the names of over 750 men killed while employed on this dangerous work. I hope in a small way this book too, will help to keep alive the memory of the brave individuals of Bomb Disposal. It is also dedicated to the memory of those who risked their lives trying to prevent the V-weapons getting through. They include the secret agents and resistance workers, the pilots who tried to shoot the weapons down (many of whom were killed as their aircraft flew through the debris from the exploding missiles) and those imprisoned by the Nazis who attempted to slow down or sabotage the production of these weapons.

Chapter 1

June 1944: The First V1s

The Allies were well aware that the Germans were developing new weapons at the Peenumünde site on the Baltic coast long before any attack on the UK materialized. Some physical evidence of this new technology even found its way over to the UK via covert means.

On 13 May 1944 a signal was received by the Air Ministry in London from the Air Attaché in Stockholm, drawing attention to Swedish press reports of a 'German radio-controlled rocket aircraft' crash at Brosarp in southern Sweden on the morning of 11 May. This was not the first report of its kind, but with growing evidence of the bomb's imminent use the British now took a greater interest than before. Two experts, a radio specialist by the name of Squadron Leader Calvert and armament specialist Flight Lieutenant Heath, were flown over to Sweden on 19 May with the intention of examining the wreckage. In the course of discussions between the two British and the Swedish technical officers, it emerged that the Swedes had previously recovered two similar projectiles, and there were two others known to have crashed on the German-occupied Danish island of Bornholm. All five had come down as part of the V1 test program.[1]

The first of these fell on Bornholm on 22 August 1943 and was investigated by Lieutenant Commander Christiansen, a Danish mine disposal officer.[2] The missile had apparently crashed 2 kilometres west-north-west of Bodilsker Church, having just missed the tops of the trees close to a house some 250 metres away. Christiansen photographed the wreckage and sent a report and drawing to the Ministry of Marine. He also sent four photographs to the intelligence section of the Naval Staff. When the Germans asked if he had taken photographs he denied it, but unfortunately they found one of the reports he had written in the possession of a 'messenger', a sailor working on the Elsinore–Helsingborg (Sweden) ferry.

On 1 September a Luftwaffe officer interviewed Christiansen, showing him one of his own photographs. During the course of questioning, a report came in of a mine in 'Hammer' harbour, and as Christiansen was the resident mine disposal expert the Germans postponed the interrogation in order for him to go and work on it. Once finished he had it in mind to make an escape to Sweden using the mine disposal craft, but two German vessels kept him under close observation the whole time.

Back at Ronne, Christiansen was interned and subjected to further interrogation, this time at the hands of the Gestapo, who charged him with espionage. In their custody he was to suffer such cruel treatment that on 8 October he had to be admitted to hospital. A couple of weeks later he was still in hospital when some of his fellow countrymen made a successful attempt to rescue him. He at last made it to Sweden, but it was March the following year before he fully recovered from his wounds.[3]

According to contemporary British records, after the Bornholm missile a second one was recovered from the sea by the Swedes (date unknown). It was given a cursory examination and then blown up as a mine. A third was also recovered from the sea, by the Swedish Navy off Karlskrona in December 1943, and on 8 April 1944, a fourth V1 crashed 20 metres from Stampere Farm, near Stamperegaarden in Ostermarie on Bornholm. The fifth was the one at Brosarp mentioned earlier. All five were believed to have been fitted with dummy warheads but from their size it was possible to determine the likely size of the charge they would be able to carry. The British experts sent to Sweden were able to return with some useful intelligence material, including a sample of the fuel used.[4]

The Germans went on with their V-weapon program and construction began on launch sites along the northern coastline of France. These did not go unnoticed. Brave men such as Michel Louis Hollard, later known as 'the man who saved London', went to great lengths to get information about these sites back to the Allies. Hollard's citation for the Distinguished Service Order in 1945 read: 'Hollard, at great personal risk, reconnoitred a number of heavily guarded V1 sites and reported on them with such clarity that models were constructed which enabled effective bombing to be carried out.' On one occasion he strolled into the construction site pushing a wheelbarrow full of bricks he had found in a ditch. Like the Dane, Christiansen, Hollard was also later caught and subjected to torture by the Gestapo.[5]

The bombing of the V1 launch sites severely delayed their construction. However, with the 6 June D-Day invasion the Germans' need to get the V1s operational became much more urgent. By 12 June preparations at some launch sites were complete and they were ready to start the first attacks on England. That evening the first salvo was fired, but the Germans only managed to launch nine V1s, all of which failed to reach England. A second salvo was launched in the early hours of the next morning. Ten V1s were launched, but again technical problems meant that only four actually reached British shores.[6]

The involvement with the flying bombs by bomb disposal personnel in the UK really began on that day, 13 June 1944, when the SR8 department of the Ministry of Supply was informed that flying bombs had actually been used for the first time against Britain.[7] Two days later a representative from SR8 went to Air Intelligence Branch 2(g) at Harrow to inspect the remains of three V1s that had exploded in open country. Not much was learnt as no fuze fragments were discovered.

On 16 June the USAAF base at Boreham, a few miles north of Chelmsford in Essex, reported an 'unusual missile' that had fallen just outside the airfield. An RAF bomb disposal officer, Flight Lieutenant Cripwell of 6210 BD Flight, based at North Weald, took a squad of men to investigate. He found what was then referred to as a 'diver' – a V1. It had exploded but bits of it were collected and kept at Boreham and Air Intelligence was advised to come and look at these as they were evidently parts of the new German 'rocket projectile'.[8]

On 17 June, lacking component parts from the actual warhead, Major John P. Hudson (Bomb Disposal), Mr Grew (Armament Research) and Robert 'Bob' Hurst (Ministry of Supply Scientific Research Dept 8), spent the entire day examining the sites where V1s had exploded, sifting the soil for any useful fragments. A number of interesting items were found, including parts of the bomb casing and an unfired electrical igniter from the elevator control gear. They noted that the craters differed, some large, some shallow, and they were of the opinion that two fuzes with differing delays were responsible.[9]

That same day an Air Ministry instruction (No 754) was issued to all RAF Bomb Disposal units. Its title was 'German Expendable Pilotless Aircraft', and it stated that sites of V1 explosions should be cordoned off and nothing removed until an investigation of wreckage had been conducted.[10]

It was actually on 19 June that the first unexploded V1 crashed to earth at around 0100 hrs, after being attacked by an aircraft. A number of V1s were shot down that night but it is believed this particular one was a victim of a No. 96 Squadron Mosquito, operating from West Malling and crewed by Flight Sergeant McLardy and Sergeant Devine.[11] The V1 came down near the south coast at Lower New Barn Farm, Fairlight, in East Sussex. Police Constable Horace Crouch was one of the first on the scene at 0400 hrs. He found a crater 5ft in diameter and 2ft deep in the corner of a field some 300 yards from the nearest building, where the flying bomb had crashed and disintegrated. The warhead had broken open and the explosive contents lay scattered over an area 100 x 50 yards. He described the rest of the wreckage as also being scattered over this area that extended into Rufters Wood.[12]

It was the following day, however, before it was reported to Hudson. He and Hurst arrived at the site around 1930 hrs and found Air Ministry Intelligence

16 Disarming Hitler's V-Weapons

The first V1 to come down in England without exploding hit Lower New Barn Farm at Fairlight in Sussex. The crash site was a little way from the farmhouse (just seen far left of top photo). A report of the time said the V1 initially impacted the corner of a field (believed to be in what is now the paddock area pictured above), with the wreckage strewn 100 yards into the adjoining Rufters Wood (right of photo). In 2011 the author and friends, with kind permission of the farm's owner and the manager of the woods, investigated the area with metal detectors. Some small corroded fragments were found, but none were confirmed to be pieces of V1. Among them was shrapnel identified as being from anti-aircraft shells. These pieces could have come from ack-ack fired at one of the many other V1s that passed this way heading for London. (*David Ransted*)

representatives already there, as well as the local Royal Engineers Bomb Disposal unit. The missile had obviously crashed at a fair speed as pieces were spread around, many having come to rest in the undergrowth. They found the bulkhead between the warhead and the fuel tank and most of the bomb casing. Some of the explosive filling also scattered around the site was collected so an analysis of its make-up could be undertaken. To the experts' eyes it looked similar to that used in the SB1000 parachute bomb known as 52A, together with a biscuit of RDX. An electrical fuze (El.A.Z. 106) and its pocket were discovered separated from the bomb, as well as another fuze-like object marked 'Ent 106'. This was found loose and had no provision for fitting a gaine (gaine being the small explosive-filled cylinder found attached to most German fuzes).[13]

The El.A.Z. 106 fuze had already been disturbed by a member of the RAF Regiment, who had picked it up. As a precaution it was given a further remote control 'jerk test' before it was considered safe enough for closer examination. The fuselage and control units were in good condition and Air Intelligence branch 2(g) took charge of these, removing them from the site at 2230 hrs that evening.

Hudson and Hurst, having completed the initial recce of their first unexploded V1, arrived back in London at 0300 hrs. Later that day the fuze (El.A.Z. 106), still in its tubular pocket, and the other fuze-like object (Ent 106), were taken to Woolwich, where scientists took a stereo-radiograph (a kind of X-ray). The Zus 40 booby-trap that was regularly found hidden in the pocket under fuzes in standard aircraft-dropped bombs was of course in the mind of the bomb disposal officer/scientist tasked with exposing this V1's secrets.

From the radiograph pictures it appeared that there was no booby trap below this particular fuze, so it was extracted from the pocket and more radiographs were taken of the now exposed fuze. By 1300 hrs the fuze and radiographs were being discussed at a meeting of interested parties at Berkeley Court, the London HQ. The electrical fuze was found to be fired by a nose rod switch which was also among the wreckage. The UXB Committee deduced from the design that this fuze was fitted in the nose of the bomb in a central axial exploder pocket.

At 1700 hrs a message came that another fuze pocket had been found at the Fairlight crash site and, contrary to orders, had been taken to the regional HQ at Tunbridge Wells. A field photography van was sent down, with Hudson and Hurst following. (Field photography was the term commonly used at the time to disguise the fact that it was a type of radiographic X-ray.) The radiographs showed a simple mechanical 'all-ways' impact fuze with a clockwork-arming device. This clockwork train was thought to be similar to the No. 67 fuze found in the infamous butterfly bombs. As it appeared to be armed, urea-formaldehyde resin was used to fill the striker chamber, as was standard practice in similarly designed bomb fuzes to prevent them from operating. They were careful not

18 Disarming Hitler's V-Weapons

Major John Hudson seen here preparing to freeze the battery in a bomb fuze using liquid oxygen, thereby rendering the fuze 'temporarily' inert. (*John Pilkington Hudson*)

to get the resin into the clockwork, as it would make it difficult to look at later. Then it was withdrawn from the pocket and the explosive gaine was removed, witnessed by the Regional Commissioner SE Region.

The fuze (now safe) and its pocket were brought back to London at 1330 hrs, and further stripping took place. This showed the fuze had actually been very close to firing, presumably disturbed during its carriage from Fairlight to Tunbridge Wells, as it was assumed it was not armed when it crashed or that would have caused it to detonate. The resin however, had made the fuze safe. It should be pointed out that a fuze with gaine attached is as deadly as a hand grenade if it detonates and a booster charge in the fuze pocket would cause an even bigger explosion. This fuze's pocket would have been fitted into the warhead from the side of the bomb in a similar way to those in aircraft-dropped bombs. In the remains of the warhead casing they found, however, provision for yet another side fuze, though none was actually found at the crash site. This was explained by local residents who said they heard an explosion about fifteen minutes after the flying bomb hit the ground. The missing fuze had probably exploded, but being detached from the main explosive filling it had failed to detonate it. A sufficient amount of the main casing was found to allow a rough reconstruction to be made.

On 21 June, Bob Hurst went to Woolwich and examined the stereo-radiographs with Ministry of Supply scientist, Dr John Dawson. The fuzes were taken to the Armament Research Department at Fort Halstead where they were photographed in detail. The radiographs and fuze components were discussed with Dr Cox (possibly Cocks) and Mr C. S. Hudson of the Electrical Engineering Dept of the Royal Aircraft Establishment, Farnborough, who had been asked to assist. They were left to trace the fuze's electrical circuits and establish exactly the method by which they functioned.

Hurst visited Fort Halstead in the late afternoon of 22 June to look at the dissected fuze. No batteries or condensers were found inside it, so it was deduced that the energy necessary to fire the igniters must therefore come from a source external to the fuze. It was clear that tracing the complete wiring system of the aircraft and establishing the source of its power was essential. The other fuze-like object (Ent 106) still had its circuitry to be established. It was apparent at this point that if an electric bomb fuze (EI.A.Z. 106) was found with all the wires to it severed, then it would be isolated from a power source and therefore 'safe' – assuming of course there was no booby-trap concealed beneath it!

24 June 1944 was an important day in the fight against Hitler's new weapon. Official records show it was on this day that another V1 was reported as having come down without exploding and this one was much more of a prize for the bomb disposal fraternity.[14] This 'buzz-bomb' was in much better condition than the Fairlight bomb. It had been shot down by a fighter plane and had come to rest at Strawberry Hill Farm, Staplecross, Sussex.

A team led by Major John P. Hudson arrived at the crash site at 0800 hrs that day. With Hudson were the two scientists from the Ministry of Supply, Dr John Dawson and Bob Hurst, with C. S. Hudson from RAE. The names of others in attendance that day included Marshall, Ballard and Newitt. John Hudson recalled in 2006 that he thought Ballard was possibly a civilian scientist from New Zealand. As for Newitt, there are a couple of possibilities – Professor D. M. Newitt of the Royal College of Science, South Kensington, had some dealings with the UXB committee, as he had been involved in tests using solvent to remove TNT from bomb casings and trepanning.[15] However, it was more likely to be a Lieutenant Clive Newitt, a very experienced BD officer who had served with the London-based No. 5 Bomb Disposal Company since the early days of the war. He had been awarded a George Medal for his brave work with numerous unexploded bombs, including one that his section had removed from Pall Mall and taken to a demolition site, where it exploded a few minutes after they had left.[16] Another bomb had been recovered from the National Gallery and taken to a less public spot nearby to be rendered safe later in the day. The bomb was unloaded from the lorry, at which point the men went for their lunch. While eating, they heard the bomb explode! Another that fell at 106 Chesterton Road, Kensington, was so badly damaged that the fuzes could not be removed. Newitt instead cut a hole by hand in the casing and steamed out the explosives, despite the fact that the bomb still had a No. 17 clockwork fuze that could have started running at the slightest knock, and a No. 50 fuze with tremblers fitted that could detonate the bomb at the merest pencil tap on the casing. In Newitt's citation for an OBE it actually states that he personally dealt with four flying bombs, and as a BD 'Intelligence Officer' he was often required to identify bombs and fuzes reported as previously unknown, or with markings obliterated or suspected to have been purposely disguised by the enemy.[17] Dangerous work indeed.

All the men present at the crash site that day would have been extremely well qualified for the job in hand. For example, Major Hudson, John Dawson and Bob Hurst had spent some time in bomb disposal, undertaking research and experimental work from their HQ in Romney Street, Westminster. A year before they were involved in discovering the workings of a new kind of booby-trapped fuze specifically designed to kill the bomb disposal officer who tried to remove it, and in the subsequent method of rendering it safe.

It is worth taking a few minutes to tell of the discovery of that particular fuze, the 'Y' fuze, as it shows something of the type of hazards men faced when taking apart the first of a new type of weapon such as the V1. It was Captain Frank Carlile who was responsible for retrieving the very first Y fuze from a 500kg bomb found buried under the track of the Bakerloo Underground Line running beneath Lords cricket ground. It appeared to be a standard impact fuze that for some reason had not functioned, the number 25b, stamped in a small circle on the face of the fuze, denoting its type. There were other markings including a letter Y, but they were thought to be manufacturer's marks of no consequence. Captain Carlile went through the normal procedure for dealing with an impact fuze of this type and pumped some fluid into it. This action discharged any electricity still held inside the fuze. He then took the bomb to a 'bomb cemetery' at Hampstead Heath, where he tried to remove the fuze with equipment that would allow him to stay out of harm's way. The 'remote' method did not get the jammed fuze out so he resorted to good old-fashioned brute force, up close and personal, using a crow bar, hammer and chisel (remember he thought he was dealing with a now inert fuze). On finally getting the fuze out he could see it was longer than usual and had a ring attached over it that made withdrawal difficult. It turned out that this was a new and previously unseen type of fuze. This particular one had a fault in its circuitry; it should have detonated the bomb when it was moved or when the fluid was pumped into it.[18] (Captain Carlile's luck was not to last. He was killed in August 1943 at his Horsham HQ, when a fuze he was handling exploded.)

Very soon the scientists developed a method for dealing with the Y fuze that involved freezing the fuze's battery with liquid oxygen so that the voltage output would go down to zero – though if the fuze was allowed to warm up while it was being removed, the battery (and the bomb) would become live again. Major Hudson pioneered the freezing technique with scientists John Dawson and Bob Hurst. They worked on live bombs to ensure their methods worked before asking other BD officers to adopt the procedure. A novel idea in regard to freezing was to stick wads of wet cotton wool on the bomb casing at intervals, away from the face of the fuze. After liquid oxygen had been poured into a clay cup stuck round the fuze, the cotton wool would freeze. It was important to check that the wad furthest from the fuze was frozen solid, to ensure that the fuze was frozen to a suitable depth too. Major Hudson's contribution was recognized and he was awarded a George Medal for his work in dealing with a bomb at a flourmill near the River Thames in the Albert Bridge Road, Battersea.[19]

Back to the V1 at Strawberry Hill Farm: John Hudson recalled in later life that the bomb had come down relatively intact in a wood, not near any buildings, but across a field or two from the road. The Ministry of Supply's UXB Committee 1944 report stated that the warhead was found intact in a wood, with side fuzes in the four o'clock position. The remains of the nose fairing and compass were close by. The rest of the V1 was a little further away (nearly a mile is mentioned in a Tunbridge Wells Home Security report), though John Hudson definitely remembers seeing the 'motor and funny little wings' during his initial investigation of the crash site.[20] Records held at the East Sussex Records Office area are a little more precise, stating that the missile was first found at Strawberry Hill Farm, map reference WR203403, though the warhead and fuel tank were missing. These were soon found a little further to the north at Wellhead Wood, map reference WR196416.[21]

The first thing the men did was to listen to the bomb with a stethoscope. It was not ticking. Hudson soon found there was a nose switch very close to the most forward point of the fairing, which would close as soon as the light alloy began to crumple, and also a pressure switch under the belly of the fairing, which would close if the bomb glided in. The Ent 106 fuze was found nearby, establishing the fact that it was attached to the warhead or nose fairing and not in the side of the fuselage. In the nose a damaged electrical fuze was shown to be the same as the El.A.Z 106 in the Fairlight bomb. The bomb was propped up with timber to prevent it from rolling as the earth around it was removed to enable a 'field photograph' to be taken.

It was established that there was no Zus 40 booby-trap behind the fuze, and that the pocket was inserted inside a larger tube (axial exploder), confirming the arrangement previously forecast. As a side note, the radioactive source for the X-ray photographs was picked up by the team in a lead-lined box from a Ministry of Supply laboratory south of Sevenoaks before they went to the crash site. The team was aware of the health hazards from radioactivity but the dangers of working on a huge, unfamiliar, unexploded warhead seemed much more immediate. The only protective clothing Hudson could remember using while handling the radioactive material was a pair of heavy gloves.[22] (Bob Hurst went on in later life to become director of the experimental nuclear reactor project at Dounreay.) The process of using X-rays to see inside bomb fuzes had already proved successful earlier in the war. Ministry of Supply scientists William Wiltshire and David Barnes worked in this field and were awarded George Medals for X-raying a mine at Petersfield, Hampshire, from 31 October to 3 November 1940.

A report dated April 1941, held in the National Archives, shows that a 500kg German bomb that had failed to explode after being dropped in Lowestoft

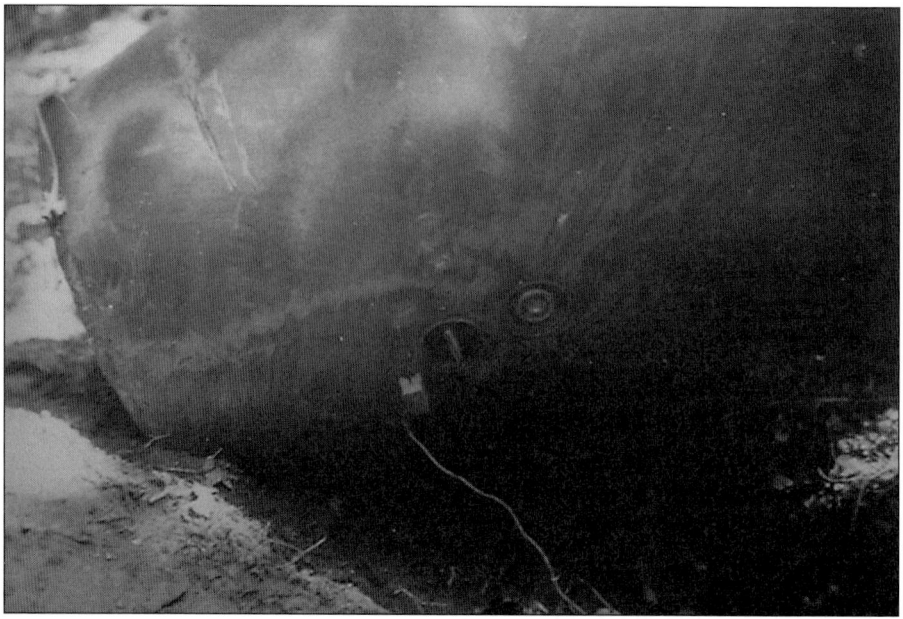

The Staplecross V1 is referred to in official records as the 'Battle' bomb (the town of Battle is about 4 miles away). The film for the X-ray was in a paper envelope in the forward side-fuze pocket. (*John Pilkington Hudson*)

The radioactive source is propped on wedges at the rear end of the bomb. It was held in a lead container with a brass screw that was removed in order to let the radon gas escape at the time of the X-ray. (*John Pilkington Hudson*)

had had its fuzes successfully X-rayed in great detail while they were still in the bomb's fuze pocket. This work had been undertaken by a Woolwich-based Research and Development team.[23]

The use of X-rays was also experimented with by RAE Farnborough back in November of 1940, when they used X-rays from a medical diagnostic 'universal' machine to actually discharge the electricity stored in the firing condensers found inside an unexploded bomb's fuze.[24] They soon found that a standard Philips industrial X-ray set was more powerful and could render a type 15 impact fuze inert in about 12 minutes by directing an intense beam of rays into it.[25] The point of the experiments was really to try to deal with the then new type 50 booby-trap fuze that had three tremblers inside, extremely sensitive to vibration or movement after the bomb had come to rest. These type 50 fuzes at first took much longer to render inert using X-rays than the type 15, about an hour in all. However, by February of 1942 they had been able to get that time down to 40 minutes. Bear in mind that as the type 50 fuzes were usually fitted in bombs that also had a 'ticking' No. 17 time-delay fuze and the 50 had to be dealt with first, time was of the essence. At the end of the day the X-ray method proved impractical. The big and clumsy (and expensive) equipment was thought too difficult to use in many situations, down the bottom of shafts or where the bomb's fuzes were lying underneath, and so on. It was also found that the equipment would have to be working at maximum output for long periods, with the possibility of overloading it to the point of tripping safety switches that could cut off the X-rays. Even more worrying for the bomb disposal officer was the fact that there was no test or way to be sure when the rays had done their job of immunizing the fuze (take the fuze out too soon and the bomb would explode).[26] The British scientists also envisaged the Germans using lead or steel shields inside the fuzes to foil their attempts at this form of immunization.

Techniques developed as the war went on. In February 1944 H. William Koch, a young physicist from the USA, arrived in Britain with a portable 4.5 Mev betatron to aid the British in taking X-rays of UXBs. This equipment had been developed by the University of Illinois under contract from the British and Bill Koch was the man entrusted with showing the boffins at the Woolwich Arsenal exactly how it worked. The X-rays from the betatrons were not only useful for seeing inside UXBs, but also in the treatment of cancer. Bill Koch had explored this idea as part of his PhD thesis. His experience with radiation led him to be employed by the Manhattan Project on his return to the USA.[27]

The V1 John Hudson and the team were faced with that day in Sussex was found to have two side fuzes in the casing. By the end of play on Saturday, 24 June 1944, the rear side fuze had been the subject of field photography and was seen to be roughly the same length as the impact fuze recovered from the Fairlight V1 – and no Zus 40 booby-trap was hiding beneath it.

Come Sunday morning a team of men descended on the crash site – Hudson, Hurst, Dawson, Newitt, Feldman (who will be mentioned later), Smith, Marshall and Karsh.[28] John Hudson remembers Karsh as a big chap from the USA, on the staff in Washington. He had met him earlier in the States when Hudson was there briefing the Americans. Yet again, field photography was the order of the day. They wanted to investigate the forward side fuze pocket and nose pocket, and also to obtain more information about the rear fuze. The photography took most of the day.

On the Monday it was just the main players in attendance – Hudson, Hurst, Dawson and Newitt, attempting to establish the exact nature of their doodlebug's rear fuze. The field photography indicated there was a clockwork time delay in the rear fuze, similar to the No. 17 fuze found in many aircraft-dropped bombs. It was therefore imperative they did not cause it to start ticking. (The clockwork mechanism in these fuzes had a tendency to restart if subjected to vibration or

A modified Philips X-ray machine being tested. The intention was to use X-rays to discharge the electricity stored in the type 50 booby-trap bomb fuze. (*Author collection*)

movement. Much like a child's clockwork toy, it might seem completely run down, but when picked up or disturbed you sometimes find there is life left in it.) More detailed photos were necessary. To do this it was decided to deal with the forward side fuze first. They would remove it and use the empty pocket to hold a specially constructed cassette for the X-ray field photo of the rear side fuze. The radioactive source for this would be placed on the other side of the rear fuze at the back of the bomb. The forward fuze had been shown by X-ray to be the same as the mechanical all-ways impact fuze previously recovered, except that the safety bolt was still in place. The starting pin had broken off on launching, leaving part of it still in place, so that the clockwork arming train had never started.[29] Field photography had also ruled out a Zus 40 booby-trap, so the fuze, now called an 80A, was removed by remote control at around 2200 hrs (it was probably pulled out with a long length of string). The arming time of this fuze was later found to be 6 minutes.

Now the rear side fuze was hopefully about to reveal its secrets! Maybe for that reason more interested parties turned up on the Tuesday morning. As well as Hudson, Hurst, Dawson and Newitt there were Greatbach, Behrendt (quite possibly Major Behrendt DCRE), Marshall and two US servicemen, Kane and Feldman. (Lieutenant Colonel Thomas J. Kane was in fact considered the founding father of the US bomb disposal organization and at the time was Director of Bomb Disposal ETOUSA. A *Newsweek* article from May 1944 described him as 'a burly Irishman with a strange air

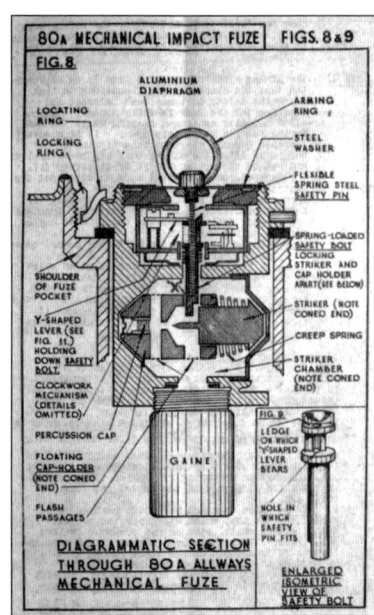

80A All-ways Fuze in author's collection. The arming pin in the top should have been pulled out by a lanyard as the V1 launched. (*Author collection*)

of indestructibility … who looks as if he could overpower single handed any bomb ever made.' Kane informed the reporter: 'In our branch of the service, you only make one mistake. You're either an expert or you're dead.' In fact Kane died a natural death in 1965. Captain John E. Feldman was his sidekick and had the nickname 'Kane's brains'.

Feldman had developed many bomb disposal tools and worked on all sorts of bombs, from a Japanese depth charge that had been accidentally dropped on a Hawaiian sugarcane field, to a naval shell dating back to Captain James Cook's days. Around the time he came to look at this V1 he could usually be found in London's Grosvenor Square, working from a shop truck converted into a mobile laboratory. A clever man, he was to be awarded an OBE by the British for his collaborative work on V1s and after the war worked his way up to the rank of lieutenant colonel, becoming skilled in the field of atomic weapons. It is thought the reason for his presence on this day was because the British Director of Bomb Disposal (1942–44), Brigadier H. H. Bateman DSO, OBE, MC, had invited Feldman to possibly try out one of his inventions. Developed in conjunction with a US Sergeant Shorling, it was called 'the Flit gun' and was a self-tapping needle capable of drilling an eighth of an inch hole in any fuze. It could make a non-leaking connection that could withstand 4000 PSI when injecting a thermosetting compound to 'jam up the works'.[30]

June 1944: The First V1s   27

Captain John E. Feldman went on to become a lieutenant colonel. (*Author collection*)

Lieutenant Colonel Thomas J. Kane went on to become a colonel and retired in 1955. (*Author collection*)

A rare photo of Captain Feldman taken a few days after he worked on the Strawberry Hill Farm V1. He is seen here working with a 'Flit gun' on a jettisoned US bomb at Shipton-on-Stour on 29 June 1944. (*Author collection*)

That morning was spent getting a remotely controlled 'Stevens Stopper' in working order. This bit of kit was used by the British BD organizations to jam a fuze by creating a vacuum inside it and then introducing a fast-setting resin into it. There was of course a danger that the fuze might detonate during this process – hence the need to operate from a safe distance. The bomb was regularly checked with a stethoscope for signs of ticking and at 1400 hrs that afternoon the cassette that had been put in the forward fuze pocket was removed and developed. It showed that no anti-withdrawal booby-trap was attached to the fuze in the rear pocket. Dawson, Kane, Feldman and Greatbach then looked into whether the jamming liquid could be impregnated with barium/bismuth in order to get a good X-ray photo of the internals. A chap named Warner was charged with finding out whether treating the liquid in this way would change its jamming properties. Warner and Feldman also had to make a reservoir from which they could measure the exact amount of jamming fluid required to fill the V1's fuze.

On Wednesday morning the Hastings High School for Girls became involved with the disposal operation, when the headmistress, Miss Commin, was asked for her help.[31] Her initial reaction to officers of the armed forces turning up on her doorstep was to worry that a soldier had got one of her girls into trouble, so she was happy simply to be asked to provide three bottles of barium and bismuth salts! It had been found that the jamming fluid mixed with barium chloride would still work, though it would take considerably longer to set. The rest of the day was spent debating the way forward and different methods to deal with the fuze, some more risky than others.

Fresh after a night's sleep, Hudson, Hurst and Dawson returned to their problem on the Thursday and took another close look at this rear fuze, even borrowing a doctor's auroscope. They were worried about introducing the jamming liquid. They didn't actually want to jam the clockwork, as they would later need to find out how long it could tick for – they just wanted to stop it from detonating. The fuze looked similar in external appearance to the forward fuze, but instead of a large hole for an arming bolt there was only a small pinhole, closed below by a black rubber disc. Major Hudson had reported to the Director of Bomb Disposal that this was an unknown type of fuze and his response was that the fuze should be recovered at all costs. (As mentioned above, the one in the Fairlight bomb had exploded 15 minutes after the warhead hit the ground, and by chance, on the same day that this V1 came down, another was reported shot down by a fighter in the Yalding district of Kent. That one also exploded 20 minutes after hitting the ground, seriously damaging nearby housing.)[32] They decided the best way forward would be to cut away the warhead's casing on either side of the rear fuze, remove some explosive and put a film holder one side, and the radio active source the other. This way they would get the clearest X-ray image possible.

A trip was made to Hastings to pick up some cutting files. After some experimenting on the casing from the Fairlight V1 they decided cutting with a hacksaw or files would cause too much vibration. It was presumed that the fuze was in an armed condition and though not 'ticking' at that moment, any vibration from a cutting device on the warhead might restart it. Then it might function after 15 minutes, 15 seconds, or instantly – nobody knew for sure. So the method chosen for cutting was by means of a nitric acid solution. (Also possibly provided by the Hastings School for Girls.) The men would have to be careful not to let the acid eat all the way through the casing to the explosives, but just to leave the metal as thin as possible, then use a very sharp knife to cut through the last part. Again a trial run was made on a piece of casing from the Fairlight V1 before working on the explosive-filled casing at Strawberry Hill Farm. This method of acid use had been developed earlier in the war and in fact Hudson had been involved in the trials, which were not without incident.

It had been at Richmond Park, on 19 June 1941, that an experiment took place in which two 250kg bomb cases were used for 'acid cutting tests'. No suitable bombs with powdered fillings were available, so Captain Hudson and a scientist, J. Gray, from the Ministry of Supply chose two empty cases from the dump and filled them with explosives taken from another bomb. The filled bombs were then separated by 200 yards and a wall of sandbags, and each was fitted with an apparatus developed by Dr T. P. Hoar of the Corrosive Laboratory at Cambridge University, under Dr W. R. Evans. In each case the acid produced from the apparatus made two small holes about three-sixteenths of an inch in diameter and the acid then proceeded to drain into the bomb itself. A bubbling was heard and the team was keen to know what was happening to the explosives inside the bomb. At great risk to themselves, they fitted a metal-cutting trepanning machine to the bomb, to cut it open and look inside. As they cut through the metal casing there was a sudden and violent eruption of yellow material in considerable quantity. Slightly taken aback by this development, Hudson and Gray, along with three Royal Engineers sappers, rescued the trepanning machine, and retired to a building further off to reassess the situation. Shortly after, at 1635 hrs on that hot sunny day, one of the bombs suddenly exploded. This was about an hour and a half after the acid had first penetrated the bomb case. Fortunately nobody was hurt – more by luck than anything else. In the following hours the other bomb's filling burst into flames without exploding, and burnt itself out.[33]

That had certainly been a close escape, but Hudson, not put off, was about to use acid again to cut into the flying bomb. He applied the acid this time using a washer bottle that sprayed a fine jet. While this was going on, at around 1120 hrs another flying bomb exploded nearby after being shot down, and later, just after 1500 hrs, they were again disturbed by a fighter firing at a V1 flying low over their heads. The threat of explosions close to where the team was working was a concern, not least because any vibration might start their clockwork fuze running.

The weather now turned a bit grim, so sandbags were put over the warhead. Hudson constantly checked with a stethoscope to see if the bomb was ticking, though this was difficult because of the constant noise from passing aircraft. Despite the interruptions, the cutting went ahead.

Next day, Friday 30 June, Hudson, Hurst, Marshall, Behrendt and Colonel G. D. De'ath (Director of Bomb Disposal 1944–46), were all on site and the acid cutting resumed. By 1300 hrs an inverted V-shape had been cut. The two diagonals were each about 12ins long, with the open end of the V straddling the fuze pocket. The triangular flap of bomb casing was then rolled back to expose the main explosive filling (it was also exposed at the front of the bomb where the cover plate had been torn off in the crash). The filling appeared as a yellow-white crystalline material (RDX/dinitrobenzene/ammonia nitrate to be precise). A rough rectangle, 6ins long by 3ins high, was also cut from below the fuze pocket in order to take the field photography cassette.

A more sophisticated X-ray was also planned, using equipment that had to be brought to the site by van, a water tank being necessary for cooling it. To this end a roadway through the woods now had to be prepared.

The explosive filling was removed by washing it out with warm acetone, assisted by gentle scraping as the surface became crumbly. The acetone was collected and reused several times over. Warm water was also used to attack the biscuit filling when a slab of this became exposed. Major Hudson and Mr Hurst started this task at 1400 hrs, taking turns to work. They stopped briefly in the early evening for sandwiches, carefully washing their hands to avoid ingesting any of the explosives or acetone that had covered their bare hands, and then continued working until 2230 hrs that night. The whole time they had to be careful not to cause any vibration. The toxic fumes produced while removing the explosives resulted in both men becoming dizzy with bad headaches – they took on an ashy pallor and their lips turned blue. Mr Smith, who lived on the farm, provided copious amounts of fresh milk that the men drank to try to counter the effects of the dinitrobenzene poisoning. Despite this Hudson spent the rest of the night vomiting and Hurst, though not actually sick, felt extremely ill.[34]

Heavy rain on Saturday, 1 July made it impracticable to get the X-ray van near the crash site. At 1230 hrs Hudson and Hurst were still feeling ill, so it

June 1944: The First V1s   31

The Director of Bomb Disposal, Colonel G. D. De'ath, visited the site of the second unexploded V1 on 30 June 1944. Meanwhile the V1 attacks continued unabated. (*Author collection*)

On the same day a V1 hit London's Aldwych killing around 46 people. Wreckage from that V1 is here being inspected on the pavement outside the main entrance to Adastral House. (*The National Archives: ref.AIR 20/6185*)

was Dawson who completed the washing away of the explosive filling below the fuze pocket and set up the field photography equipment.

During the day the site was visited by Brigadier Bateman, Marshall, Greatbach and Feather. (Feather was most likely Captain William Anderson Feather BSc, who had been awarded a George Medal back in 1941 for his work with unexploded bombs with No. 4 BD Company in some pretty important, not to mention dangerous, locations. These included UXBs next to a Royal Navy mine storage depot at Wrabness and inside the premises of an explosive and chemical products company at Harwich.)[35]

At 1800 hrs an excellent picture of the fuze was produced and a second one taken with equally good results. Half an hour later the X-ray van finally managed to negotiate the made-up road to the site, but the team figured that by the time they set up the equipment it would be too dark to work. They felt the field photographs already obtained were of sufficient quality to enable them to

32 Disarming Hitler's V-Weapons

This photograph shows the cavity that was cut above the rear side fuze pocket. The radioactive seeding is held at the top of the field photography framework. The film is not visible but is in a cavity on the opposite side of the fuze pocket. (*John Pilkington Hudson*)

carry on removing the fuze without X-rays and this task began the following day with the same people in attendance (minus Feather). The photos showed the fuze had an identical time-delay mechanism to a No. 17 fuze, as previously thought, giving a maximum time-delay of 2 hours. It was not possible to tell if this particular clock had already run and if so for how long. Because of the presence of the rubber sealing disc it was not possible to inject anything to jam up the clock's works – and as mentioned that would make later inspection difficult anyway. No booby-trap was present so the fuze would be removed, taking the chance that if the clock started in the process it could be dealt with (explosive gaine quickly removed) before it could explode and negate all their efforts. To facilitate this, cords and pulleys were attached that could jerk the fuze straight out on to a magnetic clock-stopper (used to stop ticking fuzes on other bombs). The fuze locking rings were removed using a modified Quilter key by remote control (the key had been developed earlier in the war by another BD man, Raymond Quilter, for removing German fuzes).[36] Mr Hurst sat in a trench 50 yards away listening to an electric stethoscope, while Major Hudson, about 100 yards away, manipulated the cords attached to the fuze. At 1800 hrs the fuze was pulled out and made safe. It was found that the fuze had been

June 1944: The First V1s   33

An example of a flying bomb's clock fuze in the author's collection. Known as the No. Z17Bm, this fuze contains no electrical parts but is purely mechanical. Before fitting the fuze in the V1 the armourer would adjust the time delay by turning a mechanism on the underside of the fuze, then screwing on the gaine. (*Author collection*)

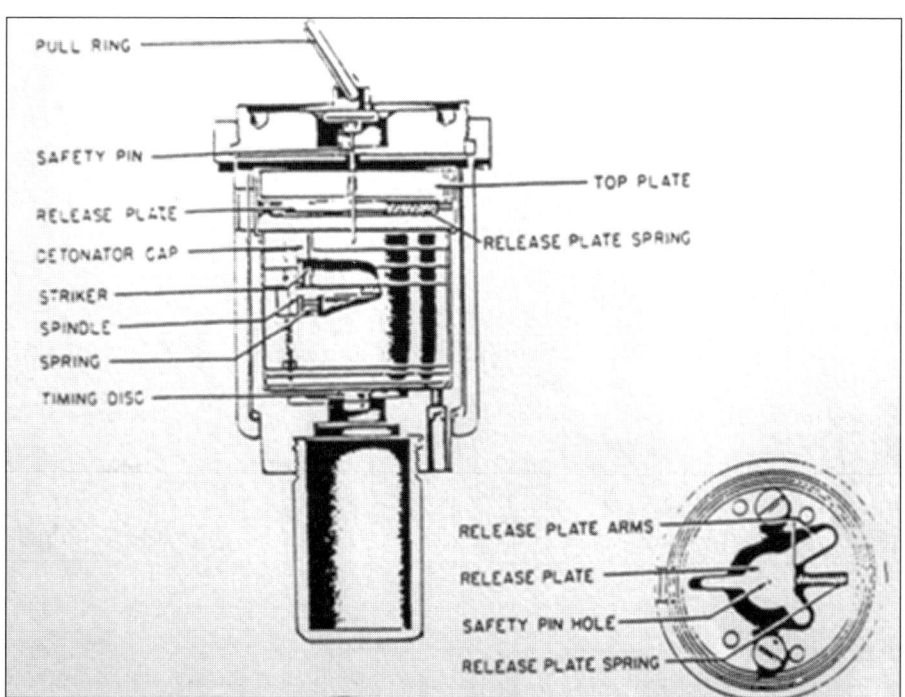

Diagram of one of the first Z17Bm fuzes recovered from a V1. (*Author collection*)

fully armed and the clock free to start running, though it had never actually started. After the explosive gaine had been removed the fuze was given a shake, which was all it took – the clockwork mechanism immediately started, and ran for 33 minutes before trying to fire the now inert fuze. On further examination the rubber sealing washer was found to have a self-sealing hole through the centre, which closed up after the withdrawal of the starting pin on launching. It was evidently designed to function as a self-destruct device, detonating the V1 shortly after its fall, should the impact fuzes fail.

The 106x fuze in its pocket was now extracted from the nose exploder tube and the remainder of the explosive in the warhead was steamed out. The dinitrobenzene fumes created again poisoned some of the Royal Engineers sappers tasked with the job. These men were probably from No. 20 BD Company based in Sussex. A personal diary of Sergeant Fred Harding from 196 Section, 20 BD Coy, recorded that sometime between 30 June and 12 July, he was tasked with collecting a quantity of explosives and a couple of detonators from a V1 at Heathfield.[37] The detonators could well have been the 'dive' ones from the tail-end (more about these later).

The completely defuzed bomb was handed over to CRE 2 Bomb Disposal Group, Royal Engineers and the intact casing sent to the RAE at Farnborough. The whole operation had lasted from 24 June to 2 July, most of the work taking place in daylight hours. At night the team had stayed at a local pub, apart from Major Hudson, who took the opportunity to stay with his mother and brother at their house in Burgess Hill for a taste of 'domestic normality'. He recalled it felt a bit odd, especially as his mother did not know what he was up to, though she knew it involved bomb disposal – to use his words 'a weird situation'.[38] For their efforts, both Hurst and Dawson were awarded George Medals and Major Hudson received a bar to add to the medal he had won the previous year. Bob Hurst, a New Zealander by birth, was later put into uniform and went to Germany at the end of the war to work on Allied unexploded bombs. It was not his first taste of Nazi Germany. While in New Zealand as an undergraduate in the 1930s, he belonged to a group of students who helped Jews to escape from the Nazis. One of those he helped was the philosopher Karl Popper, whom he got to know quite well.[39]

Around this time, late June 1944, Home Security were issuing instructions to the Civil Defence forces that they should report on whether any traces of a wireless unit were found among V1 wreckage. It was thought that the Germans might be using transmitters to plot the course of individual V1s.[40]

The Allies now had gathered information on how the V1s worked. In summary they had three fuzes:

June 1944: The First V1s

Bob Hurst photographed post-war. (Dounreay News, *May 2012*)

Hudson on the day he received his medal. (*John Pilkington Hudson*)

A piece of the Strawberry Hill Farm V1 still exists. For some years the wing spar sat in the farmyard. When the owner of the farm sold up he donated the spar to the Robertsbridge Aviation Society and it can now be seen in their museum. (Photo: David Ransted)

1. The El.A.Z 106 in a fuze pocket at the forward end of the exploder tube with a sensitive pressure-plate nose switch. This functioned in the case of nose impact. An electrical switch mounted on the forward underside of the fuselage would detonate it in the event of a belly landing. There was also an inertia bolt within the El.A.Z 106 just to make sure! (An important point

36 Disarming Hitler's V-Weapons

to bomb disposal personnel was the fact that if the wires to the El.A.Z 106 were cut, it would be rendered harmless, whereas cutting the wires to the pressure switch would detonate it.)
2. The forward side fuze pocket contained an 80A mechanical all-ways impact fuze. The all-ways fuze would operate in whatever attitude the bomb landed, in circumstances where the El.A.Z 106 might fail. This fuze was particularly sensitive to jarring. Tests proved that dropping on to a hard surface from a distance of five-eighths of an inch could cause it to function.
3. Rear side fuze pocket contained either an 80A fuze or mechanical clockwork delay fuze.

Debris from doodlebugs was still collected for the authorities to scrutinize. In a number of cases documents were found amongst the wreckage – technical papers that, for whatever reason, were left inside the missile when it was made. These obviously were very useful to the British, not only for technical information, but if dated it would indicate the lapse of time between final inspection at the factory and launch. The type of documents found included

This photo was used on the front page of an American newspaper, *The Evening Sentinel*, on 23 June 1944, with the caption: 'British pilots examining a flying bomb that failed to explode when it fell.' Identifying its location is a little difficult. One would assume that this photo, released on 22 June, would show the Fairlight V1, as that was the only recorded unexploded flying bomb up to that time. That bomb completely broke up though, which doesn't look to be the case here. This wreckage looks very similar to one in the introduction to this book, which was reproduced in an intelligence report dated 22 June. (*Author collection*)

June 1944: The First V1s   37

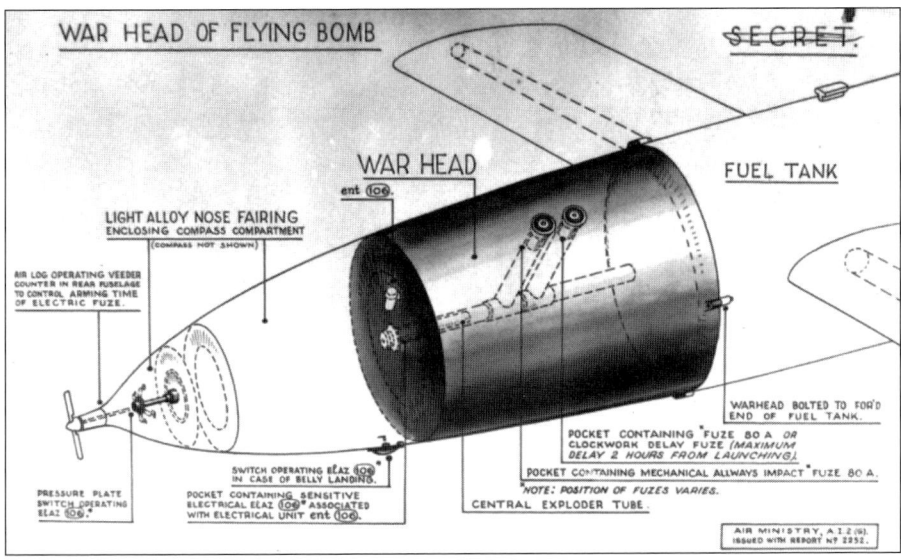

As can be seen by the description of the V1's multiple fuzing system above, the Germans really wanted these bombs to go off! (*The National Archives: ref.HO 199/465*)

test data for the barometric height control unit, rate of turn gyros and the automatic pilot as a whole, as well as a job sheet for the main spar.[41]

As a result of collecting debris for analysis, sometimes a V1 was reported as unexploded because such large pieces were found, including the engine unit or wing spars. There is a signal for example, held at the National Archives, from Home Security and copied to a number of addressees, including the Director of Bomb Disposal, that states that on 27 June 1944, at 0030 hrs, a suspected unexploded flying bomb had come down at Peabody Buildings,

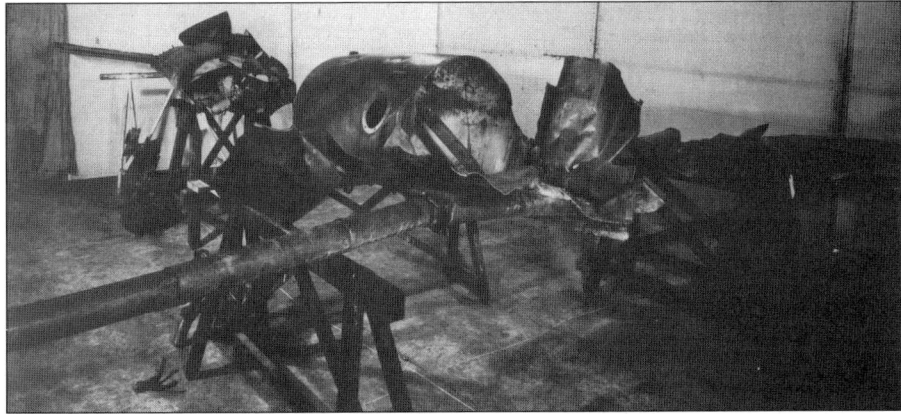

Wreckage of a V1 at RAE Farnborough. (*Author collection*)

Drury Lane, in London's West End. However, this information was corrected by a pencil note to the effect that it had 'now exploded'.[42]

At the same time the Strawberry Hill Farm V1 was being worked on there were reports of others coming down unexploded. On 25 June 1944, a fighter brought down a V1 almost intact in the Snargate area of Kent – possibly subject of a Home Security signal of 4 July, that an 'unexploded flying bomb complete' had come down at Brookland (near Snargate) – time of falling unknown.[43] And on the 27th two more reports came in. One missile was brought down by a fighter in the Horsmonden area of Kent, and the records actually state 'UXB? with wings and jet unit complete. One seriously injured.' The other was reported to have come down unexploded at Wormdale Farm, Newington, in Kent. Tom Ledger, whose parents ran the farm during the war, was born in 1943 and was never told of any unexploded V1s having landed there, although his parents had said that a V1 exploded in an apple orchard about 400 yards south-west of the farmstead. Some damage was caused to an outbuilding and the windows of the farmhouse facing the bomb were blown out.[44]

Around 0940 hrs on 29 June 1944 a V1 was caught by a barrage balloon of 945/7 Squadron at cable site 527. It crashed unexploded in a cornfield half a mile south-east of Downe, Kent. (The site was about 50 yards from the western boundary of the field to the west of Hang Grove Wood and about 250 yards south-east of Downe Court.)[45] The bomb was buried in the ground in an upright position so that the rear end was just level with the ground. The rest of the V1, including the jet propulsion unit, was lying on the surface a few yards to the east.

The spot where a V1 came down, photographed in 2010. Downe Court is to the right. The doodlebug hit the ground without exploding, some 50 yards in front of the line of trees. (*Author collection*)

A Colonel Biggs joined Major Hudson at the crash site of the Downe V1. He is seen (above left) removing the storage caps from the fuzes. On launching, the Germans omitted to remove these caps and attach the lanyard to pull out the pins that armed the fuzes – hence it never went off. Colonel Biggs is hard to identify from the photo but is believed to be Alfred John Biggs from Neath (above right) who distinguished himself earlier in the war when attached to 9 Bomb Disposal Company in Birmingham. He was an experienced bomb disposal man, awarded the George Medal for working on a 250kg bomb that fell through a railway viaduct at Curzon Street, Birmingham, in November 1940. On that occasion the fuze was ticking and was jammed in its pocket. Biggs had to resort to a hammer and chisel to take off the filling cap at the back of the bomb and washed out some of the TNT with a hose. He then used a crowbar to lever out the fuze pocket. The fuze exploded 7 minutes after its removal from the bomb. (*John Pilkington Hudson and Lionel Meynell*)

It was noted at the time that a serial number, 1/4303, was painted on the upper surface of the tail. The two fuzes proved to be mechanical impact fuzes with their storage caps still in place. They had never been armed. The complete arrangements for starting the fuze and the number 80A were thus disclosed. Field photography was employed to check for booby-traps. None were present so the fuzes were removed.

The nose fuze was, as before, an El.A.Z 106. The Ent 106 was found to be damaged, and it could be seen that it was fixed to the filling cover plate very close to the El.A.Z 106. The explosives were steamed out – despite full protective clothing being worn this time, including respirators, the operators still suffered minor effects from the fumes.[46] The intact case was sent to the USA.[47]

Frame set for an exposure on the forward side fuze. Notice the stencilling on the warhead '52A' denoting the explosive filling used. (*John Pilkington Hudson*)

Major Hudson seen here fixing a fuze extractor. (*John Pilkington Hudson*)

June 1944: The First V1s   41

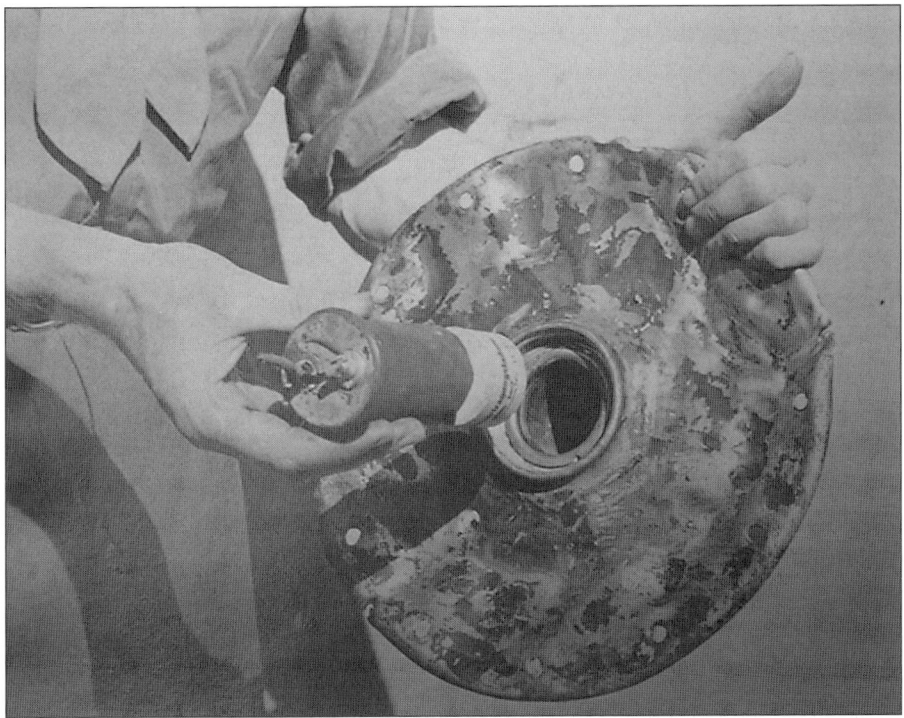

EI.A.Z. 106 nose pocket from the Downe bomb. Notice the white picric ring that fitted over the explosive gaine on the end of the fuze. (*John Pilkington Hudson*)

Radiograph showing right angle views of the fuze pocket from the Downe bomb containing the 106x fuze. (*John Pilkington Hudson*)

Chapter 2

More V1 'Duds'

On the 29th of June, 1944, yet another V1 was reported as having come down unexploded, this time on open ground at Benenden, Kent, at around 1230 hrs.¹ With the V1 attacks continuing, the Bomb Disposal HQ staff were eager to disseminate information to their officers as quickly as possible on how V1s could be rendered safe. On 1 July, an informal 'secret' information bulletin called *Fuze News* gave some basic details. It referred to 'the pilotless aircraft, also known as the Flying Bomb, Buzz Bomb, Doodlebug, and a selected nomenclature unsuited for printing'! It went on to detail the 'elaborate' fuzing system with some diagrams.

People with 'a need to know' were also made privy to technical information about V1s. For example, Edward Carter, chief warden for Waltham Holy Cross in Essex, was invited to the Bomb Disposal Training Wing at Chelsea on 23 August 1944, to look over the remains of a V1. He found it to be badly damaged as a result of its crash landing, roughly finished, and a lot larger than he had expected. That night he went home to see more of these weapons flying over his area.²

The armed forces were by this time doing their best to shoot down the V1s before they got anywhere near densely populated areas, by firing at them from the ground, from aircraft or from ships in the English Channel – an example being on the night of 5/6 July, when over the Channel a Typhoon from 137 Squadron, based at Manston, intercepted two V1s. In the unit's war diary F. O. Nicholls describes how after attacking them he saw one explode as it hit the water, while the other just made a splash, indicating that it had not detonated. Its remains could well still be lying a few miles off Boulogne, beneath one of the busiest shipping lanes in the world.³

On 8 July 1944, a flying bomb exploded at Tellis Coppice, near Battle in Sussex. The wings and nose of the missile were recovered and what was thought to be one unexploded fuze. An RAF expert, Flight Lieutenant Ella, who was actually an intelligence officer based with No. 49 Maintenance Unit at Faygate, was called in to look at it. He found it was only an electrical switching component.⁴ The maintenance unit had the task of recovering crashed aircraft including V1s (records show they collected the wreckage of

13 V1s during June and July 1944), so it made sense to have an intelligence officer attached to them to check on anything found that might be of interest.[5]

These unexploded V1s did not always come to rest in the relatively deserted countryside. On Sunday, 9 July 1944, at 1440 hrs, a V1 struck and demolished the single-storey back addition to a house in Northumberland Avenue, Welling, Kent (official records say it hit number 234, though one eye-witness says it was number 236).[6] Doreen Ashmeade lived at number 208. Over sixty years later she still remembered the events of that day:

> I was an 'old' ten and a half-year-old girl when a doodlebug came down in our road and did not explode that Sunday in July. Our road runs parallel to the railway line from Welling and Falconwood stations. The German bombers seem to follow this line onwards to London, also the doodlebugs too. As children we found the war time an exciting time. When bombs fell on our concrete roads the surface was replaced with asphalt which was very good for rollerskating.
>
> On this Sunday afternoon my friend (Hazel Bullard who I am still in contact with) and I were rollerskating in the road, when we heard the noise of a Doodlebug coming towards us. We knew from previous doodlebugs that had passed by that if the engine was still making a noise we were safe. BUT suddenly the noise stopped and we heard instead the rush of wind as it came towards us. Hazel and I rushed for home. We just made it in doors when there was an almighty crashing noise … then silence … the bomb had not gone off.
>
> When we went outside, I can't remember how long we stayed indoors, but we were told that the Man (owner of the house) was sitting in the kitchen eating his late lunch. I was told or heard that he liked a pint on his way home on Sundays. His wife and child were under the stairs in a cupboard and were protected in some way. They were not hurt.
>
> The bomb went through the front or side of the house into the kitchen under the table. I was told the man ran along the doodlebug to turn off the gas and water which was making a hissing on the V1. This all must have happened when we were still indoors.
>
> Very soon afterwards officials or wardens came to every door and made us leave our homes and we had to leave our doors and windows open. Looters were around at some time while we were away.
>
> Our household was full as we had three grandparents who had moved in with us due to their own homes being bomb damaged, my parents, and three sisters. We had a table shelter in the house and a garden shelter which slept six. We were sent to 'The Green', a small shopping area ten

Wreckage from the Northumberland Avenue V1. Notice the round compressed air tank. (*After the Battle*)

minutes away, where we had our food in a recreation hall, which was above the shops. People had to sleep in disused garden shelters. The one we slept in was damp and had no door, it was horrible as I recall. We had to take our animals with us and I took my cat but it disappeared during the night and was never seen again. My sister who was out rambling on the day returned at 6pm but was not allowed to go home, she had to travel to London next day to work still wearing her grubby shorts and shirt – she was most upset. Hazel, my friend, thought we were only away one night but I thought it was three.[7]

The V1 Doreen had witnessed broke up completely on impact and did quite a lot of damage despite not exploding. A number of the local children, including 11-year-old Ray Scott who lived just down the road at number 227, tried to get a better view of the incident and hopefully obtain a souvenir. The authorities were quick to prevent such activities. However, on his return from the Co-op hall at 'The Green', to which local folk had been evacuated, he did see the compressed air spheres awaiting collection at the side of the road.[8] As the crash site was just 50–75 yards from the railway line all the trains had to be stopped.

The missile had broken up to such an extent that the explosive filling was widely scattered. The nose fuze pocket was recovered, with its El.A.Z 106 damaged and with no gaine present in the gaine case (perhaps sabotage at the factory where slave labour was employed). Both side fuze pockets were recovered. One contained an 80A fuze in which the cap had fired but again there was no gaine present![9] The other pocket had its complement of picric acid but no fuze fitted. From later evidence it is probable that this pocket did not always contain a fuze but had the space filled with cardboard held in place by a bakelite disc. The area was declared safe and the next day the local residents were allowed to return to their houses. What remained of the Welling V1 and its warhead were recovered by the Royal Engineers from the property's back garden.

Another V1 that came down in a populated area fell in the middle of the night at Malthouse Road, opposite Crawley railway station, Sussex, on 29 July 1944, requiring the evacuation of many residents. The local report centre received a message from Malthouse Road that 'something was falling from

Royal Engineers removing the V1's remains. (*After the Battle*)

the air, and it was hissing'. Thirty minutes later another report came in to say that something had fallen in the road, and was still hissing! Apparently the hissing was caused by compressed air escaping from a damaged air bottle in the relatively intact unexploded V1.[10] A UXB Committee report mentions a V1 that was shot down in Crawley (possibly on 31 July), with its warhead still attached.[11] A Home Security report also says a V1 came down unexploded in Crawley on 3 August and 120 people were evacuated.[12] The bomb was rendered safe on 4 August. These dates obviously do not all tally, but most likely all refer to the Malthouse Road bomb (contradictory reports are not uncommon). According to local resident Rex Robinson, the V1 actually came down on 29 July. His wife's cousin was born just a couple of days before and was one of those who had to be evacuated. He said the baby was named Hazel Elliot and went through life being known as HE, for 'High Explosive'! Rex recalled that the V1 had landed in the allotments behind Malthouse Road at the Southgate end.[13]

In a UXB committee report the fuzing of this V1 was described as 'normal' – one El.A.Z. 106 in the nose and two 80A fuzes in the side, both armed. In one case the striker had just pricked the cap but not fired it. This indicated that the glide landing had not been sufficient to operate the mechanical impact fuze. The explosive filling now known as 52A+ was steamed out of the warhead which was then passed to the Air Ministry.[14] Later in the war a Bomb Disposal Bulletin, 12 July 1945, would warn against steaming out V1 warheads, as by this time a number of them had been found to contain propellant mixed with the main filling, rendering the steaming-out process dangerous. It noted that the RAF had reported that one had detonated during the operation.

The knowledge and parts being acquired by the bomb disposal fraternity around this time were very quickly shared with the intelligence organizations and the Special Operations Executive were given details of the warhead and its firing system. SOE had some interesting ideas for combatting the threat, using their agents on the Continent. An SOE 'Crossbow' progress report of 10 July 1944 stated: 'It is thought possible to manufacture a number of specially designed fuzes having an external appearance identical with those fitted by the Germans. These fuzes would be designed to effect detonation at the moment when the flying bomb leaves its launching ramp.'[15] Basically as the safety pin was withdrawn on the launch ramp the warhead would detonate.

At a meeting at SOE's Baker Street HQ a few days later a hastily manufactured counterfeit 80A 'All-ways' fuze was passed round and it was agreed that 100 of them should be produced by the 'next moon' period for delivery to the Continent. The secret gadget workshop at Station XII was to

Numbers 234 and 236 Northumberland Avenue no longer exist. They had stood on this spot until an unexploded V1 smashed into them. The new houses seen here in 2010 were built on the site but now form part of the adjoining Millbrook Avenue. (*Author collection*)

The same day as the V1 crashed unexploded into Northumberland Avenue, pieces of another flying bomb recovered by the men of bomb disposal were undergoing trials at Swynnerton, near Stone in Staffordshire. The aim of the trial was to find which ammunition was most effective and the components that were most vulnerable. The trial also attempted to find the optimum angle to shoot from in order to bring them down or explode them in the air. (*Author collection*)

lead on the manufacture, the first fake fuze being produced and tested by the end of the first week in August. All were to be completed by the end of the month.[16] Whether any of these were successfully mixed with the real fuzes at the factories or while the component parts were in transit, is not known.

Back in the UK, Home Security Tunbridge Wells reported on 21 July that a flying bomb had exploded 16 minutes after impact at Icklesham in Kent, only a mile or so from where the first unexploded one had come down at Fairlight a month before.[17] There was no mention of casualties – surprisingly, as people's natural reaction tends to be to rush to the scene of where something has crashed.

Quite a few flying bombs were failing to get through the defences. One shot down on 23 July into the Channel, by Flight Sergeant Vassie (a Spitfire pilot of No. 1 Squadron based at Lympne), was seen to cartwheel across the water before sinking.[18] The cartwheeling action meant the 'all-ways' fuze was obviously not functioning as it should have – and this is another case where the rusty remains of a V1 may still be sitting at the bottom of the Channel, 10 miles south of Dover.

Early the same day another doodlebug was shot down by a fighter and crashed unexploded at Scrums Farm, Sandhurst, Kent, just after 0600 hrs.[19] It was most likely shot down by a Tempest from 3 Squadron, based 10 miles away at Newchurch. The warhead fell some distance from the fuselage and buried itself 12 feet into soft mud. A number of people were evacuated from their homes nearby.

On the 25th the warhead was uncovered. Just to make things awkward the fuzes were underneath. The nose fuze was as before, an El.A.Z. 106, but the Ent 106 could not be found. The forward side fuze pocket had a white circle painted around it and the fuze appeared to be a normal 80A mechanical impact type that had been armed. The rear side fuze pocket had a red circle painted round it and contained an armed 80A fuze that had been damaged. The aluminium diaphragm and washer from its top were missing. The bomb was covered in muck so it was first cleaned up. Both fuzes were washed out thoroughly, using a jet of water applied under pressure (pressure was created with a bicycle pump, using a self-tapping spigot as a nozzle – possibly the self-tapping screw of Feldman's referred to in the Strawberry Hill V1 incident). It was found that the arming clock in the forward fuze was slightly different in design to ones seen previously. As it was important to ascertain the arming time, it was decided to remove the clock without jamming it up. The aluminium diaphragm and washer (which were already partly off) were levered off and the closing plate unscrewed. The clock, complete in its case, then fell out. The rear fuze was tackled next. It had sheared off below the shoulder and the body of the fuze had settled down into the pocket by about ¾ in. It was stuck fast in the pocket and would not come out. This fuze was 'jammed up' and arrangements were made for field photography. The casing was cut around the rear pocket with a hacksaw, lubricated with water. Tin snips were then used to complete a square cut which detached the fuze pocket from the main body of the bomb. The metal was then cut back close to the fuze pocket with the tin snips. Acetone was then poured on to the explosive filling around the pocket where it rapidly made a grey sludge, releasing the stiction of the explosive on the pocket. After a time the pocket was levered out. The locking and locating rings were removed and an American fuze head attachment for a hydraulic clock stopper was inserted, to serve as 'anchor' for the levers.

The fuze pocket and fuze were then destroyed in a controlled explosion. Later investigation showed that both the 80A fuzes had functioned, but had failed to ignite the gaine detonators. It was thought that the soft mud had absorbed a lot of the impact so there was not enough force to operate the fuzes in the normal way. Instead, the sudden ingress of mud into the bomb had forced the fuze's strikers into their caps, but by this time there was so much mud inside that the flash was absorbed and the main filling failed to explode.[20] The filling was found to be of a different chemical composition – cast trialen '106'. (This aluminized TNT/RDX filling was a much more powerful explosive. It was estimated that by using this the area of blast damage could be increased by 70–80 per cent.) The intact case was later sent off to Woolwich to be used in static detonation trials to learn more about this explosive filling.[21]

There was actually a need for further warheads for experimental purposes, so using those found in the unexploded V1s the Ministry of Supply had blueprints drawn up and from them the Ranalah Company in Lombard Road, Merton, South London, secretly produced a replica warhead. It was delivered to the Woolwich Arsenal a couple of days after Christmas 1944, at a cost of £54. Another company, Shalson Manufacturing Co. of Brettenham Road, Edmonton, made replica fuze pockets and associated fittings in early 1945. These would later be destroyed in 'blast' trials. British Intelligence found that as the Germans were suffering from a lack of raw materials late in the war, a couple of V1s used against targets on the Continent actually had their warheads made from plywood. They also found out that the Germans had wanted to use Myrol liquid high explosive in the warheads, but that would have been unsuitable for the plywood. Instead they investigated the possibility of producing them from resin-impregnated paper.[22]

Late in the evening on 27 July it was reported that a fighter had brought down an unexploded V1 at Knock Wood, near St Michael, Kent.[23] On this

(Top) An unexploded V1 at an unidentified location. In the lower photograph notice the frame attached to the warhead that holds the radioactive material used when taking X-rays. (*Author collection*)

day over 30 V1s were intercepted by Allied aircraft. It was the first time that the British put a jet fighter up against the V1s. Meteors of 616 Squadron based at Manston flew patrols, although they were unlucky in their attempts to shoot anything down. One pilot had trouble with his guns just as he had a flying bomb in his sights. Another closed in on a V1 but had to break off the attack because of the approaching danger of barrage balloons.[24] Many other types of aircraft flying that day did have better luck, including Tempests from 150 Wing, 3, 56 and 486 RNZAF Squadrons, Spitfires from 41, 91 and 165 Squadron, Mustangs of 129, 306 and 315 (Polish) Squadron, and later in the evening a Mosquito of 25 Squadron.[25]

Despite all the attacks by aircraft, the anti-aircraft batteries' actions and the barrage balloon defences, the flying bombs were still managing to get through.

At lunchtime the next day, 28 July, yet another V1 was reported to have come down unexploded in Kent. Ten-year-old Ian Doyle, then living in Tunbridge Wells, recalled the bomb. At the time he was walking through a recreation ground with friends and they heard the sound of a V1 engine. Then the sound stopped. Immediately they took cover by a wall and waited for the impact, but there was no explosion so they continued on.[26] Another schoolboy, Bob Tollett, was at home when he also saw this V1. It was being pursued by a fighter but it disappeared into a cloud. The next thing he saw was the V1 coming across the playing field at a height of around 20 feet, heading for his house! He ran down into the basement shelter but heard no explosion.[27]

The V1 in question had actually been brought down by the fighter and had crash-landed in a built-up area of Southborough, just north of Tunbridge Wells. Another witness was 15-year-old Les Chapman. He saw the missile flying erratically and a policeman who had been seeing some children across the road to what was then Christchurch School, shouted a warning to him to take cover. At that moment the V1 levelled out a little before crashing on to an allotment belonging to Mr Alfred Medhurst, in Powder Mill Lane. He was the caretaker of St Matthew's School, a little further up the road.[28]

One account says that it slid along, losing its nose fairing in the process, but still stayed relatively intact, finishing up just a few feet from the wall of a house with the bomb still attached to the fuselage. As it skidded along the ground it allegedly ploughed through some barbed wire that caught an unfortunate man tending his allotment and swept him along with it and he had to be extricated by those who rushed to the scene.

James Richardson, another local resident at the time, recalled the event. He was at home when he heard the engine of the V1 stop, but no explosion. He went out the back of his house, looked along the row, and to his horror saw it gliding towards him just a few feet above the rooftops. He dashed back into the family's Morrison shelter but still didn't hear an explosion. He recalled that some time later the local air raid wardens came round the streets with loud hailers, telling people to open their windows because there was an unexploded bomb nearby. James and some friends went out and found the V1 had landed in a smallholding, not far from James's home. It was not barbed wire that it had become tangled up in, but chicken wire. James was told it was a man tending his chickens (rather than his allotment) who was also caught up.[29] Les Chapman also remembered being told there were chickens running over the V1 as people drew near it. Some were not so quick to approach the scene – according to Les, a bus driver who had seen it come down was reversing his bus at speed in the opposite direction.

A couple of youngsters had a narrow escape. Bert Clinch, who was eight at the time, was picking blackberries with his cousin Bob only 20 yards from where the bomb came to rest. They saw it in the last seconds before it hit, as it flew along Southview Road with its engine spluttering. It narrowly missed the chimneys on Southview House (the first house on the corner of Southview Road and Powder Mill Lane) and then crossed Powder Mill Lane before hitting an apple tree on the caretaker's allotment. Bert's cousin Bob made a dash for cover and dived through a downstairs sash window of Southview House. Two elderly ladies happened to be in the room taking afternoon tea at the time. Bert, however, was not so quick. He stayed where he was, looked around the hedgerow and saw the V1 had come to a stop in the caretaker's chicken run. To him it appeared to be sitting 'in' the apple tree, that had split as a result of the impact. Over 60 years later he still clearly remembers the sight of the V1's red hot exhaust and hearing short bursts of its motor as it tried to restart (the design of the engine was that a spark plug was used for the initial start, but after that the heat of the engine would ignite the fuel/air mixture as it entered the engine. This it did in pulses – hence the characteristic sound of the 'pulse-jet' engine). Bert didn't see anyone with the V1 but he was himself caught in wire – not chicken or barbed wire, but the telegraph wires that had been severed as the V1 passed by. The blackberries were abandoned as he freed himself and ran for home. A Mr Goodyear, who ran a tea shop in Powder Mill Lane where the bus crews took their breaks, took part of the apple tree away later and cut it into rings. He drilled a hole in them and made them into medallions with red, white and blue ribbons, and on them he inscribed something like, 'This tree lost its life but saved the lives of many others'. These were then sold off for the war effort. Bert's father kept his for many years.[30]

Hundreds of people had to be evacuated from their homes because of this V1 – some going to Nell's Bridge, a brick-built cattle bridge at Mr Hetterley's Farm, 2 miles away.[31] Royal Engineers bomb disposal officer, Lieutenant Eric Sivil, with some of his men from No. 20 BD Company was tasked with rendering this V1 safe.[32] One of the men who would work on recovering this

More V1 'Duds' 53

Pictured in 2011, the spot where a V1 fell without exploding 67 years before. Bert Clinch was standing roughly where the sign post is when the V1 narrowly missed the chimney of Southview House (far right), snapped the telegraph lines and landed close to where the tree now stands (far left). (*David Ransted*)

Lieutenant E. W. Sivil (left) and Major C. R. Hardham had the job of dealing with the Southborough V1. Sivil not only survived his time as a BD officer, but lived to an old age, passing away in January 2006. (*Lt Col Eric Wakeling Ret'd*)

bomb was Walter Clary, who in later life went on to serve the same community in Southborough as a Labour councillor.[33] Bert Clinch remembered that the bomb disposal unit were billeted at what is now a garage, on the opposite side of Powder Mill Lane to where the doodlebug came down.[34]

This particular V1 was the first one Lieutenant Sivil had ever been required to deal with, though he was no stranger to the dangers of bomb disposal work. On this occasion Sivil's CO, Major C. R. Hardham also attended, lending his authority to the operation. The two men checked that the evacuated area was sufficient and looked over the wrecked V1 lying on the surface. Any markings were noted down. In this case the fuze pockets had red and white circles painted round them. A couple of fuzes were quickly identified, the nose fuze being an El.A.Z., and an 80A was seen to be fitted in the rear side fuze pocket. The front side fuze pocket had a bakelite cap over it devoid of any markings – something that had not been seen in the previous unexploded V1s. Field photography equipment had been sent for, but on inspecting the bomb Sivil decided that it was in too sensitive a condition to fit the equipment without risk of it going off. The armed 80A fuze had some fast-setting resin injected into it to jam up the mechanism.[35]

This being a high profile incident affecting many people, the Bomb Disposal Group Commander, Lieutenant Colonel R. O. St J. Marshall, also arrived to add his support. The police were urging a quick resolution to the situation, and Marshall and Hardham discussed the course of action. Though they were still waiting for the field photography to arrive, they decided that Sivil should press on without it. He withdrew the jammed-up 80A and turned his attention to the fuze pocket covered by the bakelite cap. He was of the opinion that there was no fuze fitted and that this was a kind of storage cap, but without X-raying the pocket first there was no way of being certain of what was hidden beneath. He held the cap down with one hand as he removed the locking ring holding it with the other, just in case there was a spring-loaded booby-trap beneath the cap. No pressure was felt so he removed the cap and found a roll of corrugated cardboard acting as a spacer – as he had suspected, no fuze was present. The third central fuze pocket was tackled next. This also presented something of a challenge for Sivil. There were wires connecting the fuze with batteries and an airscrew generator. These had to be identified, cut and taped up in order to render the fuze safe. Sivil got a couple of his men, Sappers Brooks and Evans, to help him with this and then he removed the fuze.[36] This accomplished, the front end of the V1 had been made safe. However, in the back of the missile were two more small detonators. Once these had been removed this particular warhead, which still contained the explosive, was passed to the Air Ministry to enable their experts to take a closer look and was also used in static detonation trials.[37]

The two detonators referred to at the back of the V1 were part of the mechanism that put the missile into a dive. Contrary to popular belief, the V1s were not designed to simply run out of fuel at their target; instead a counter

More V1 'Duds'   55

A number of V1s came down without exploding as a result of being attacked by Allied aircraft. Flight Sergeant D. J. Mackerras, based at Newchurch, Kent, was responsible for destroying 14 V1s (3 of those shared with others). On 6 August 1944, while attempting to tip a V1 using his Tempest aircraft's wingtip, he lost control. The Tempest went into a spin from which he was unable to recover and he was killed in the crash. Mackerras was only 23. He is buried at Brookwood Cemetery in Surrey. (*David Ransted*)

Anti-aircraft fire and barrage balloons successfully brought down many V1s. This substantial surviving piece of wreckage is the wing from a V1 brought down by a barrage balloon at Hadlow, in Kent. It was kept as a souvenir by the captain of a local anti-aircraft battery and is now on display with a number of other V-weapon relics at the Wings Museum near Balcombe, West Sussex. (*Author collection*)

calculated the distance travelled. At a predetermined distance these small detonators at the back of the V1 exploded, with the result that the compressed air lines operating the control surfaces were severed. The rocket's elevator locked in a neutral position, the control from the gyro to the rudder was cut, keeping it in a neutral position, and two spoilers dropped from the underside of the tail plane. The spoilers put the bomb into a dive and the engine then cut out due to fuel starvation – the angle of the dive could prevent fuel from being sucked up by the engine. Not only that, the rocket relied on compressed air to mix with the fuel, and the air lines had been cut, allowing the air to escape. Sometimes there were malfunctions, possibly due to anti-aircraft damage, and engines cut only to restart again during their descent. Occasionally these dive detonators were not fitted at all (sabotage, human error or lack of parts?) and in those cases the bomb glided down at a shallow angle as the fuel ran out. Though the 'dive' detonators fitted were only small, they could still be lethal close up. If they had not been fired in flight then they would have to be made safe on the ground by the BD officer. This would be quite a common occurrence for V1s that were shot down, as the mechanism that recorded the length of flight, driven by a small propeller at the front of the missile, would not have reached its preset distance. An example of this was a V1 shot down by ack-ack that exploded at Lodge Field, coincidentally at another Powdermill Lane, near Battle in Sussex, on 20 July 1944. The two rear detonators were removed from the wreckage by members of the Royal Artillery and taken to Battle Police Station for subsequent collection by a bomb disposal unit.[38] The police station was actually slightly damaged by another V1, again shot down by ack-ack and exploding close by at Stream Farm, on 4 August. The two rear detonators from that bomb were also found and brought to the station.[39]

In fact a few V1s were known to have yet another explosive charge on board. It was found that some had a 'propaganda leaflet ejector' fitted. This was a tube that extended out the rear of the fuselage below the engine. On launching, a safety fuze was ignited that burned during the flight and, at a predetermined time, ignited a bag of gunpowder that blew the leaflets out.

On 29 July, Home Security Tunbridge Wells reported that a suspected unexploded V1 had been shot down by a fighter at 2310 hrs, a mile or so north-west of Tonbridge, at Hildenborough.[40] This is most likely the V1 that Spitfire pilot Flight Lieutenant Bruce Moffett brought down. He reported first sighting the V1 lit up by searchlights 15 miles south-east of Tunbridge Wells. He 'chased it to Tonbridge' where he shot it down at 2305 hrs. However, he said that the V1 exploded on the ground. Unless it was just fuel he saw exploding, then the warhead of this one can be discounted as an unexploded example.

In August 1944 more V1s were reported as having come down without exploding. Shortly before midnight on the 3rd, a V1 was attacked approximately 18 miles south-east of Dover by Flight Lieutenant Brandreth flying a Typhoon. The war diary for 137 Squadron describes how Brandreth closed to 250 yards and fired a short burst of cannon. Strikes were seen on the starboard wing and the engine and as a result the V1 dived into the sea but did not explode.[41]

Of the V1s that made it across the Channel, one was reported to have come down unexploded in the Hawkinge area of Kent at The Firs Farm, on the Elham Road, on 10 August.[42] This one is thought to have been shot down by Pilot Officer Edwards, in a Mustang of 129 Squadron, who reported it as spinning down after he shot at it, a mile east of Lympne.[43] (An official report talks of an unexploded V1 on 14 August that came down in the Hawkinge area. It broke up, the report noting that its two 80A fuzes had no gaines attached.)[44]

Then at noon on the 11th, a farm labourer discovered a relatively complete V1 that had landed in an old crater made by a bomb previously dropped in 1940. It was on the edge of a wheatfield against Chase Wood, Frant, approximately a mile WNW of Frant Station. The warhead was missing, but there was no outstanding report of an explosion that had not already been verified and satisfactorily cleared up. It was therefore thought that the warhead was somewhere nearby, hidden in thick undergrowth or woodland. Amazingly the occupants of Brook Farm, some 250 yards from where the V1 was found, were completely oblivious to the fact that a bomb had fallen nearby. The labourer who found it reported to the local police, who in turn told their HQ at 1430 hrs. By 1730 hrs that day, a representative from Air Intelligence Branch A.I.2.g, Flight Lieutenant Rudd, visited the site. As some parts of the wreckage were rusty, and rusty rainwater sat in other parts, it was thought by the police that this V1 had actually come down some three weeks before it was discovered. A report at the time stated that 'it would need an army of searchers to comb the surrounding district' for the elusive warhead. Another report concluded that this particular V1 (serial number 215082), could be discredited as a UXB.[45] It was believed to have been shot down by an aircraft only three days before (not three weeks) and though it does not say so in the report, it is assumed that there must have been some evidence discovered that it did in fact explode.[46] Possibly it broke up in the air and the warhead exploded in a remote spot some distance from the main wreckage.

An Air Ministry Weekly Intelligence Summary for the week ending 12 August 1944 stated that 6 unexploded V1 warheads had been recovered.[47] It was noted that different explosive fillings were used: 4 contained 52A+ filling (Matrix: RDX, dinitro-benzine, ammonium nitrate 17/53/30. Biscuit:

RDX, ammonium nitrate, calcium nitrate 21/48/31); another was filled with Trialen 105 (RDX, TNT, aluminium 15/70/15); and the last one contained Trialen 106/109 (Matrix: RDX, TNT, aluminium 25/50/25. Biscuit: RDX, aluminium, wax, 70/25/5). It was assessed that the Trialen 106/109 probably gave the heaviest blast wave of any filling at that time employed by the enemy and that the aluminized fillings accounted for abnormal amounts of damage reported at certain incidents.

A few days later, on the 16th, another buzz-bomb is reported as having come down just before 0900 hrs, in a field one third of a mile south-east of the junction of Lodge Lane and Addington Village Road, Croydon.[48] This V1 was reported to have gone into a shallow dive and skidded across a field at Pledge Farm, then went through a hedge, narrowly missing a tractor. The warhead apparently split open without exploding and was later found to have a damaged EI.A.Z. 106 and one 80A fuze that had run for a minute, and another 80A that had run for 5 minutes.

At around 0420 hrs on 30 September 1944, yet another unexploded flying bomb was reported as having come down at Nazeing in Essex.[49] This was the first example of an unexploded V1 that had been 'air-launched'. As the Allies pushed through Europe after D-Day, the Germans lost many of

Looking south down Lodge Lane, Croydon, in 2010 (notice where the tram lines now run). It was the fields on the right where the unexploded V1 slid to a halt, on 16 August 1944, having punched through a hedgerow (almost certainly the one pictured here). (*Author collection*)

their V1 launch sites. As an alternative, the Luftwaffe had He111 aircraft converted to carry V1s. Obviously these could be launched from different directions at different targets, increasing the range of the V1 over its ramp-launched counterparts. During this night's attack, 8 V1s made it to the UK (11 launches were aborted and one was shot down by anti-aircraft fire over the sea).[50] Of those 8 that made it over the coast, 2 were shot down, including the one that came down at Nazeing without exploding.[51] When the wreckage was examined it was found that no detonators to activate the spoilers had been fitted to this particular example.

These bombs would break apart quite easily on hitting the ground, assuming of course they were not totally destroyed in exploding, as was normally the case. The scattered fragments could sometimes be found in quite a dangerous state, as happened at Hopton in Suffolk, on 15 October 1944. Here another air-launched V1 was hit either by anti-aircraft fire or in an attack by a fighter plane. In the book *Air-Launched Doodlebugs – The Forgotten Campaign*, it states that this particular V1 glided down near Back Lane, Hopton, in Suffolk. The missile apparently crashed through a hedge and broke up. The crash was heard by Mr Pitcher of Bali Cottage and reported to the authorities.[52] Though parts of the wreckage were found intact (engine, compressed air tank and mainspar), the warhead had broken up without exploding. The remains were investigated by three Royal Engineers bomb disposal officers – Major Thomas J. Deane, Captain Yard and Lieutenant Charles J. Bassett.[53]

These three were all well aware of the dangers of BD work and knew to expect the unexpected. Back in January 1941, Deane had experienced a nerve-wracking time with a 1000kg bomb at Great Baddow, Essex. During the process of having its explosives steamed out, the extension cap fitted to the fuze suddenly blew off. Despite not knowing the reason for this small explosion, Deane stayed with the bomb and carried on with the steaming out. He had been awarded a George Medal in 1942 for this incident, as well as for working on a number of bombs during the danger period immediately after they had dropped and when they were most likely to explode should they be fitted with delayed action fuzes (this was done in the dark and where the bombs had come to rest, in a railway gas works and petrol depot in the Saxmundham area of Essex in 1941).[54] Apart from defuzing numerous bombs on his own, recovered from such places as a lily pond in Leiston Agricultural College, and one from under a footpath at a church in Lowestoft (that one had required a pneumatic drill to be used to break the concrete above the bomb, which could well have been fitted with a fuze sensitive to vibration), he had also helped the Royal Navy in dealing with quite a few unexploded parachute mines, so you could say his credentials for working with this V1 were pretty good.[55]

The V1 had come down in a field and bits of it were spread over a wide area, including the surrounding hedges. The warhead had broken up in such a way that the fuze pockets, complete with their booster charges, had separated from the main explosive charge. As a precaution the officers temporarily moved the fuzes so that they were even further away from the main charge. While the men continued to investigate the site, one of these fuzes suddenly exploded with great force. Lieutenant Basset was killed and Yard and Deane were injured in the explosion.[56] What was left of this flying bomb was removed by the RAF's No. 54 Maintenance Unit.[57] Within a few days Captain Yard was back at work, dealing with clearing our own minefields around Lowestoft. Major Deane on the other hand had to spend some time in hospital recovering.

At around 2020 hrs on 5 November 1944, another air-launched V1 was shot down by ack-ack in the county of Suffolk. It actually fell on an anti-aircraft gun site at Gorse Hill, Aldeburgh, approximately quarter of a mile north of Crag Pit Farm, landing on Number 3 gun, on site S7 of 438 Battery, of 138 Heavy Anti-Aircraft Regiment.[58] The fuel tank split open on impact and a fire ensued. One gunner was severely burned and three others injured. The warhead broke away and fell on to the gun-laying radar equipment, where it broke into pieces. It was fortunate that it never exploded as ammunition for the guns was stacked nearby. A couple of observers from a nearby ROC post witnessed the rocket pass within 10 feet of their position with its engine still running, and reported that the gunners' tents had caught fire.[59]

Captain Yard of 10 BD Company, pictured here, was wounded along with his colleague, Major Deane, when a fuze from an unexploded V1 detonated as they were inspecting the wreckage. Another officer, Lieutenant Bassett, was killed. (*Lt Col Eric Wakeling Ret'd*)

On the night of 11/12 December there was another attack using air-launched V1s – a total of 14 rockets were to be launched from Heinkels, but 7 launches were aborted. One was shot down into the sea and a couple more were victims of ack-ack.[60] One was reported to have come down near the coast, approximately half a mile west of Hopton/Corton coast, Suffolk, at

2243 hrs. This V1 broke up on impact, about three-quarters of a mile northeast of Whitehouse Farm.⁶¹ It left a trail of debris scattered over an area of 100 yards, which included the explosives from the torn open warhead.⁶² Captain Yard, who had been wounded a month earlier when he inspected V1 wreckage in similar circumstances, cautiously dealt with the debris. The loose explosives were collected up and destroyed in the depression made by the crash.

The new year of 1945 saw the attacks by V-weapons continue. On 3 January an air-launched V1 crashed at Neaves Farm, Stalham, Norfolk. It went into a ditch, burying its fuselage and wings, the warhead disappearing into the soft wet earth of the side of the dyke.⁶³ Initially the authorities could not get anywhere near the spot as burning fuel in the area prevented any close examination.⁶⁴ On 14 January, after unsuccessful attempts to recover the warhead, it was blown up where it lay, but in 1990 relics from this V1 were found and put on display in the Norfolk and Suffolk Aviation Museum.

On 16 January 1945, three bomb disposal officers, Captain Herbert James Hunt, Captain G. Tyson and Captain G. F. Horsfall met at CRE HQ Kensington to be briefed about an unexploded V1.⁶⁵ They were informed that it had fallen in Woodbridge in Suffolk, 3 days before, next to an airfield used by the RAF for emergency landings (map ref 87/818672).⁶⁶ Another air-launched V1, this one apparently had been shot down by ack-ack fired by the 7th City of London Regiment stationed in the area.

A relic from the Neaves Farm V1 on display at the Aviation Museum at Bungay, Norfolk. This is the slat from the underside of the fuselage for the launch piston to hook on to. (*Author collection*)

The men were told the bomb should be X-rayed before taking the fuzes out, to be sure of their type, and to check for booby-traps. Horsfall was something of an expert with field photography and had been given the nod the night before, so he had the opportunity to pick up the necessary equipment from Kent early that morning in his specially converted van. His driver was a staff mechanic who would be able to assist with the work. The three men set off for Suffolk in their own vehicles, but stopped en route for refreshments together, over which they could discuss their plan of attack. Horsfall let them know how he wanted to play it. All of them were a little suspicious of this V1, aware that the Germans were known deliberately to present 'unexploded' bombs to the British, fitted with a new type of booby-trap to kill any expert who tried to dismantle it. On arriving at Capel St Andrew, close to the crash site, the three officers were guided through some woodland to Capel Green Farm, which bordered on the airfield approximately half a mile west of Butley Abbey.[67] Here the V1 had made a belly landing in the front garden and skidded into a cesspool. All that could now be seen was the pulse jet unit sticking out of the 'matter' – the soft landing probably helped to account for its non-explosion.

In attendance was Lieutenant Barker, from No. 10 Bomb Disposal Company based at Thorpeness, Suffolk. There being no choice, the next step was to get into the cesspool and try to expose enough of the V1 to take X-rays. The men soon found themselves up to their waists, their gumboots frequently being pulled off by the suction. They found that the fuze in the nose fairing had smashed and exposing the side fuzes proved to be very difficult. Like digging in wet sand, as soon as matter was cleared away it was replaced by more. For some considerable time they tried to remove enough cess material to expose the bomb but their efforts proved futile – you might say they had a bit of a job on their hands. The only logical thing to do was to tow it out of the muck, despite the risk of detonation. The RAF next door were asked for assistance but that would take a few hours to materialize, so in the meantime the men went for a walk and sat and watched damaged aircraft that were landing at the adjacent airfield.

Eventually an RAF corporal arrived with a Cletrac (a caterpillar-tracked vehicle) and length of steel cable. The cable was secured to the tail of the V1, then laid out so that it traversed some trees, and attached to the tractor unit that had been positioned behind the farmhouse. The building would afford an element of protection should the V1 explode as it was moved. The slack on the cable was taken up and at that point the men lay on the ground next to the tractor unit as the cable took the strain. However, the flying bomb was stuck so firmly that the tractor just dug up the ground without actually moving it. The next option was to try a running tug in the hope of breaking the suction

Major Hunt inspecting parts from a V1 that exploded in The Avenue, Wood Green, London, in July 1944. His miniature medals are now in the author's collection. (*Author collection*)

of the cess material. This was risky – jolting an unexploded bomb is never a particularly good idea. However, the V1 still did not budge and the cable snapped. Another cable, thicker and longer, was sent for and another attempt made. This time they were successful and the V1 emerged from the muck. The fuzes were cleaned up and found to both be type 80s. A fair bit of time was then spent attaching the various clamps to the bomb that would hold the X-ray equipment in exactly the right positions.

Apart from working on an unexploded V1 in a cess pit, Major Hunt had to deal with other unexploded bombs in particularly nasty locations, including a 1,800kg bomb that had penetrated the gasometer in Southend Road, East Ham, London – seen here in 2002 in its deflated position. Major Hunt was apparently more scared of climbing the vertical ladder through the middle of the gasometer than of the bomb itself. (*Author collection*)

As the men had not had anything to eat or drink since arriving at the crash site and it was by now almost dark, they sent a driver to get some tea. However, when he eventually returned he was carrying the tea in a bucket – it was cold and undrinkable. Horsfall had at this point taken a number of X-rays but it was too dark for him to be certain of the results or to tell if there were any modifications to the fuzes. It was decided they would call it a day at the crash site and take the X-ray equipment back to Lieutenant Barker's base at Thorpeness, where the results could be studied under better lighting. The men did their best to remove the muck from their uniforms and clean up, though uncomfortably none of them had brought a change of underwear. They then had a meal at the base's mess and went into Saxmundham to find a pub and forget their troubles for the

A rare photo of one of the men, Captain G. F. Horsfall, who worked on the unexploded V1 at Capel St Andrew (he is on the left of the photo, which is unfortunately very poor quality having been reproduced from an unidentified newspaper clipping). He is holding the broken fuze from a large 'Hermann' bomb that had buried itself in Haling Road Timber Yard, South Croydon, in 1941. Captain Horsfall and Lieutenant G. A. Frake (right) recovered that particular bomb in December 1945. (*Author collection*)

evening. After spending the night back at the Thorpeness base, they had another look at the fuzing system and were at last confident of what they were dealing with.

Next, the fuzes were removed from the V1 without incident and on the morning of 18 January a 'bigwig', Brigadier Gilbert Streeten, came to look it over. He was the Chief Engineer Eastern Command, whose duties included visiting the sites where men were working on UXBs in order to see that everything was progressing satisfactorily. After seeing the Capel V1 he went off to meet Major Deane, now recovered from his hospitalization the previous autumn (see above), to look at mines to be cleared from a Felixstowe beach.[68]

Around this time Lieutenant Colonel G. D. De'ath, Director Bomb Disposal Home Forces, sent a memo to various BD companies, warning that ETOUSA had reported that flying bombs in the Western European theatre had been found to be fitted with a delayed action fuze that in one case exploded 2 hours 40 minutes after impact (earlier fuzes had a designed delay of 2 hours from launching).[69] Bomb disposal officers always had to be vigilant for changes in design in the weapons they were dealing with.

It is not known if the Capel St Andrew V1 was the last unexploded flying bomb to land in the UK. In Major Hartley's book *Unexploded Bomb*, published in the late 1950s, passing mention is made of an air-launched V1 that was recovered unexploded in Cheshire and another in the Manchester area.[70] Regardless of where the last one fell, suspected unexploded examples continued to be reported long after hostilities ceased.

Chapter 3

Unexploded V2s in England

As in the case of the V1s, the sites where V2s had come down, both on the Continent and in England, were investigated in order to find shrapnel and wreckage that could be exploited.

On the afternoon of 13 June 1944, several people in a small Swedish village in the area of Knivingaryd, Backbebo (22 miles north-west of Kalmak, Småland) heard three or four bangs. This was immediately followed by a number of pieces of metal falling to earth. A V2 had apparently exploded in the air. Another explosion occurred on the ground producing a crater approximately 13ft wide and 5ft deep. It was believed this was caused by the warhead though it was thought doubtful that it contained a full charge.[1] This V2 was not a run of the mill type. It was in fact fired as part of an experiment in radio-controlled steering from the ground using 'joysticks'. The small charge in the warhead was probably there to destroy the electrical equipment on board should things go wrong. They did indeed go wrong as the ground controller lost control of the rocket and the Germans did not know where it went. Also, due to what was almost certainly a fuel tank explosion, the electrical equipment fell separately from the warhead.

Two tons of material from this rocket was collected up, though the locals did manage to keep a few souvenirs. The Swedish Defence Aeronautical Experimental Institute at Bromma Airport in Stockholm undertook their own investigations over the next six weeks or so. Then, after a request from the British, the wreckage was loaded into twelve crates and flown by an American C-47 to Scotland and then on to RAE Farnborough for the British boffins to examine.[2]

Rockets fired from the trial range at Blizna and wreckage in the Sarnaki area of Poland also came under the scrutiny of the Allied intelligence network. After the firing of each projectile, German troops were detailed to search the area to recover the pieces, but some of these pieces were either photographed or spirited away before the Germans got to them. Some were to end up at RAE Farnborough.[3]

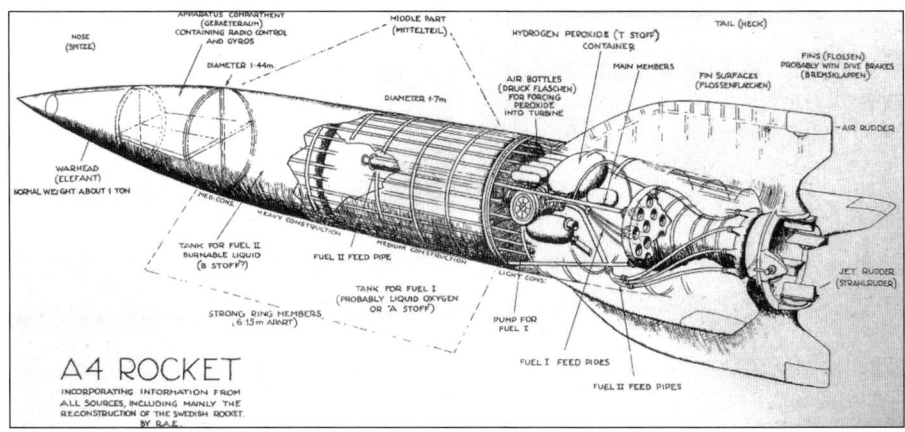

Through pieces recovered on the Continent, the Allies began to build up a picture of how the V2 was constructed and of the threat it posed. (*After the Battle*)

The first instances of V2s hitting the UK occurred on 8 September 1944. Within a very short time the British authorities were studying the sites of detonation. One of the first was at Rye Hill/Thornwood Common area, not far from the airfield at North Weald in Essex. The particular V2 that came down here actually hit just a few minutes after the very first one (which was at 1844 hrs on the 8th, at Staveley Road, in Chiswick, West London). The next day, Edward Carter, Chief Warden for Waltham Holy Cross, went to investigate the explosion he had heard the night before. He was directed to Thornwood Common and to a lane off to the west. As he crossed a ploughed field he met up with Captain Martin, of Royal Engineers Bomb Disposal, who was also investigating the incident. Martin was apparently, and not surprisingly, reluctant to commit himself as to what he thought had caused the explosion. The impact crater was found in a small wood and measured some 8ft deep and 20ft across with all the trees and shrubs blown away from the immediate area. Carter noted at the time: 'No large lumps anywhere, but lots and lots of quite small fragments of metal. Some of the metal was of the alloy type found in the parachute mine, others were quite thin, and again another was thin sheet steel. There was a certain amount of heavy stuff resembling bomb casing, but not a lot of it. Clearly not a pilotless aircraft nor Compo, nor ordinary bomb. Must be something out of the ordinary to attract all the 'big noises'. An odd feature is that no-one seems to have heard it coming.'[4]

Some residents of Sussex heard their first V2 as it exploded prematurely at a very high altitude. An eye-witness reported that the explosion was heard high in the air over the Mark Cross neighbourhood, following which a puff of smoke was seen. Then there was a rushing noise and another explosion which was, no doubt, the detonation of the warhead. Some say they heard three explosions. Some idea of the great height at which the missile disintegrated may be gathered from the fact that small pieces of metal were reported to have been falling in the neighbourhood for up to ten minutes afterwards. The casualties from this rocket were nine rabbits. Wreckage that was considered very useful for intelligence purposes was gathered up by the authorities.[5]

Another impact site looked at was at Magdalene Laver near Chipping Ongar in Essex. Flight Lieutenant Clark of the Air Ministry HQ BD staff arrived at 1300 hrs on 11 September 1944, some three and a half hours after the V2 had exploded, only to find some RAF officers from North Weald and a Captain Norris from the Royal Engineers BD based at Harlow had already investigated.[6]

Flight Lieutenant Clark searched an area of approximately 400 yards radius but found few 'working' parts. He had been looking for any radio-type parts but in his report to Wing Commander Harrison GC, said he found only a cluster of laminated tinfoil leaves, possibly insulated by thin leaves of mica. Numerous portions of insulated wire were found in the crater. He also noted that, despite the fact that some parts had been exposed for several hours, they were much colder than the soil temperature. A civilian who had touched some of the parts shortly after the missile fell found that some were too cold to handle. Specimens recovered were sent to RAE Farnborough for further analysis.[7]

As well as schoolboys wanting to get souvenirs from the V1 and V2s, so too did many adults – an added complication for the authorities, who were trying to gain as much information as possible from the fragments found after an explosion. The RAF's No. 54 Maintenance Unit, which normally had the task of recovering crashed aircraft, also involved themselves in the collection of V-weapon material for this purpose. A note from their CO to the Royal Aircraft Establishment Farnborough, dated 10 October 1944, makes mention of the remains of 'enemy projectiles' recovered from Surlingham, Rockland St Mary and Little Plumstead, Norfolk, and how some of the parts had gone missing, no doubt taken as souvenirs. Parts recovered from exploded V-weapons were even removed from an RAF lorry while it was parked at the USAAF B-24 airfield at Rackheath. This theft was reported to the USAAF provost marshall at the base. The civil police reported that certain electrical components and a compressed-air bottle that fell through the roof of The Star public house, at Rockland St Mary, were also removed by American servicemen who were believed to have come from Seething.

There was some danger associated even with the wreckage from exploded V2s. A black egg-shaped tank, approximately 34 x 20ins, contained a colourless liquid which was actually highly concentrated hydrogen peroxide, with very different properties from the ordinary household or commercial hydrogen peroxide. It could set fire to most organic material such as wood, grass, leather, articles of clothing and so on. In contact with the skin it produced white blisters, which could be painful for a time but apparently left no lasting effects. A note to the director of BD from Air Intelligence Branch 2(g) of the RAF, on 14 September 1944, suggested the very environmentally unfriendly method of leaving the tank in a convenient stream or pond to flush out the dangerous liquid.[8] The same note mentions that a deep purple liquid (concentrated calcium permanganate) might be found in the wreckage. Though not as dangerous, it advised that it should be poured on the ground before removing equipment.[9]

With more attacks occurring, the public were beginning to be better informed about the V2. Censorship was relaxed a little and soon press articles, such as in the *Daily Herald*, 11 November 1944, were providing quite a few facts relating to the collection of wreckage. That particular article talked of 'Back-Room Boys, who delved down into the crater with bomb-

The dangerous hydrogen peroxide tank from a V2. (*www.V2rocket.com*)

disposal squads and recovered twisted metal'. The writer gave quite detailed descriptions:

> Each component I have seen has been wired delicately and intricately – even fine-spun glass has gone into the lining of some of the units. Aluminium, copper and the finest steel have gone into the manufacture of V2. Tubing has been recovered ten feet in length and scraps of metal suggesting the rocket has an outer casing of thin sheeting. One part is pear shaped. It contains a liquid smelling of sulphuric acid which burns and chars anything it is applied to – it will strip the flesh from a man's hand. Another unit is frosted when found. It contains liquid air.

Occasionally V-weapons were incorrectly reported as unexploded, probably due to unfamiliarity with the weapon on the part of members of the public or civil defence organizations. On 20 November 1944, an unexploded missile was said to have landed at Fairmead Bottom in Epping Forest, Essex. Chief Warden Edward Carter was sent to check it out and described it 'as if H. G Wells's story of *The War of the Worlds* had come true, and here was a missile from Mars! In the dim light of our torches we could make out that its apparent length was about six feet, with more out of sight where it was buried in the earth at the side of the roadway.' What he was looking at was actually the combustion chamber from an exploded V2. He went on to note: 'Although it was over an hour since the missile fell, the metal was too hot to touch with the bare hand, and the heat inside almost overpowering.'[10]

A report dated 9 December 1944, by Tunbridge Wells Home Security, stated that a suspected unexploded V2 had come down at Sheepwash Farm, Cowbeech, in the parish of Hurstmonceaux. The date and time of fall was not known, but there was a vertical hole 13ft wide and over 33ft deep. A bomb disposal squad was tasked with digging the rocket out. It is not known what they eventually found, but a note pinned to the report stated that the BD team 'were still digging to find out what this could be, though it was not believed to be a long range rocket'.[11]

In fact, there were very few unexploded V2s in the UK. The first recorded instance of one was on 11 March 1945. This rocket had been launched that morning from the Statenkwartier neighbourhood in The Hague.[12] It was soon after 0700 hrs when the missile broke apart over Foulness Island, the warhead falling in a meadow approximately 500 yards west of Stannetts Farm at Paglesham in Essex.

A newspaper clipping from 10 November 1944. (*Author collection*)

First thing the next day Major Lewis Gerhold of No. 22 BD Company Royal Engineers visited the site. The entry hole was some 5 ½ ft across and about 18ft deep. Gerhold was lowered down by rope and found minute traces of grey paint stuck to the sides of the shaft. At the bottom among loose soil were pieces of plywood and fibreglass (a glass wool material was used as insulation between the fuselage skin and fuel tanks). A party of Royal Engineers proceeded to excavate the site. Though the warhead was not considered to be a particular threat to people or property, being buried out in a field, the authorities were still keen to recover it for intelligence gathering purposes.[13]

While the digging was going on another unexploded V2 was reported not far away at map reference WM088124, which is at Hutton in Essex.[14] This one had come down at around 0640 hrs on 18 March, having been launched from the Haagse Bos area of The Hague. Amazingly it hit the ground horizontally and is thought to have been 'tumbling' during its final descent, which greatly reduced its speed. That day the site was visited by Air Technical Intelligence

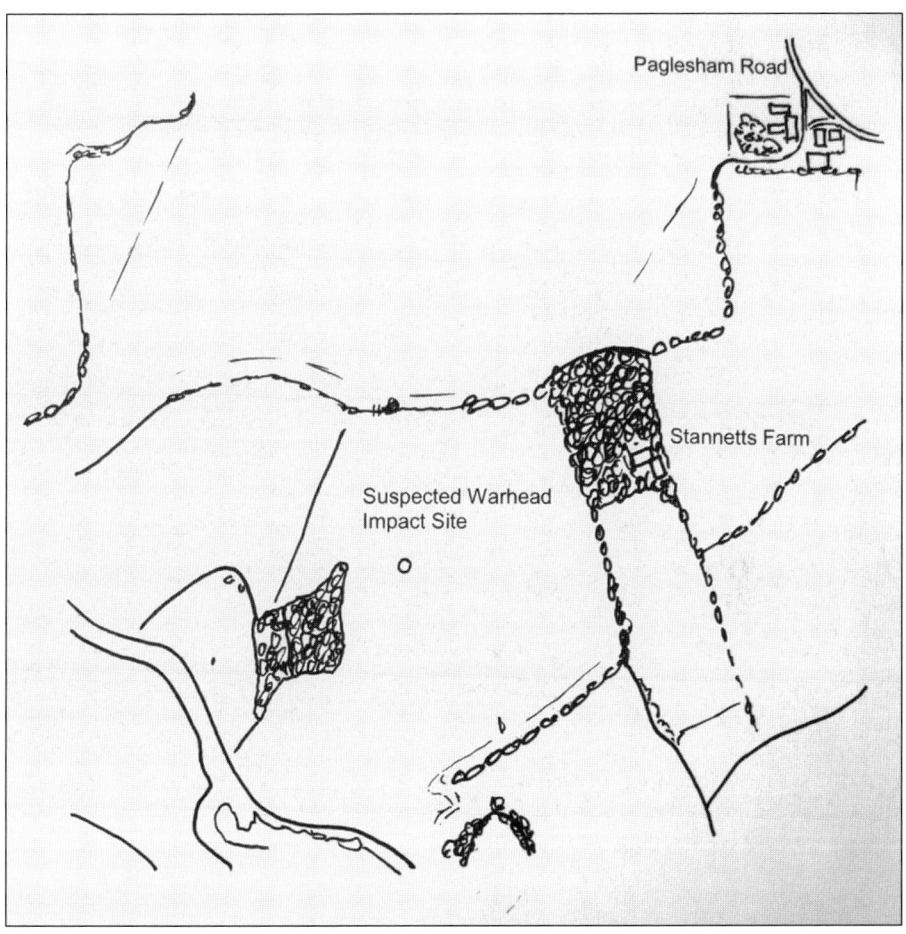

On an Ordnance Survey map there is a small pond marked in the field about 500 yards to the west of Stannetts Farm, in Essex. On a Google satellite photo the pond looks more like a muddy depression in the ground. This is believed to be the in-filled excavation where the warhead was recovered. (*Author collection*)

officers and it was confirmed to be an unexploded long-range rocket, complete in every respect but damaged by the impact. Major Gerhold left the V2 at Paglesham and went to this latest incident. He found the rocket was lying horizontally in a field in an east–west direction (tail to the west).

It was located approximately 200 yards ESE of Creaseys Farm, Hall Green Lane/Church Lane, Hutton. The main body of the fuselage had split badly from the control compartment to the thrust frame that held the engine's turbine pump assembly. A large quantity of the glass wool insulation was scattered around. It was noted that the fuselage skin covering the main thrust

Detailed images show the wreckage of an unexploded V2 at Creaseys Farm, Hutton, in Essex. (*The National Archives: ref. AIR 2/9224*)

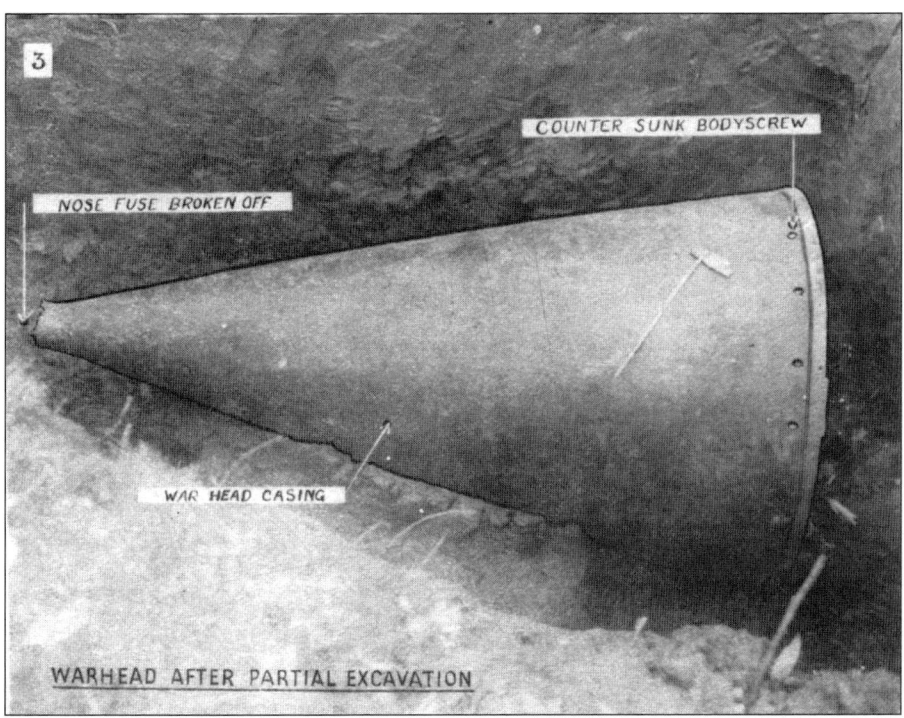

WARHEAD AFTER PARTIAL EXCAVATION

frame showed considerable signs of heating. Two of the stabilizing fins were still in good condition, the other two on the underside of the rocket had been crushed in the impact. The warhead, virtually intact, had embedded itself some 3ft into the soft clay soil. The nose fuze had broken off at the steel locking ring and was lying by the warhead. The rear fuze was intact and was connected to a complicated piece of electrical equipment called the 'sterg unit'. This was in turn connected to a power source in the form of batteries located in the control compartment.[15]

The crashed V2 attracted a lot of attention and many people gathered to look at it. To quote Gerhold, 'all interested parties bar King Farouk' were there, including representatives of the General Staff, the RAF, Ministry of Home Security and so on. Gerhold made himself a little unpopular by insisting the area was cleared and proper safety precautions observed.[16] One of those in attendance was Lieutenant Colonel S. C. Lynn OBE.

Lynn helped Gerhold convince onlookers that it was unsafe to stay there. The area was cordoned off and military guards posted. The job that faced Major Gerhold was a tricky one. The fuzes fitted, both in the nose and a secondary back-up one at the rear of the warhead, were connected to a power source via the sterg unit, located behind the warhead. In fact wires from this unit ran through the warhead's explosive filling.

76 Disarming Hitler's V-Weapons

Lieutenant Colonel S. C. Lynn was one of those in attendance at Hutton. (*Author collection*)

Glass cover at the tip of the V2's nose that protected the impact switch seen inside. (*The National Archives: Ref. AVIA 6/25653*)

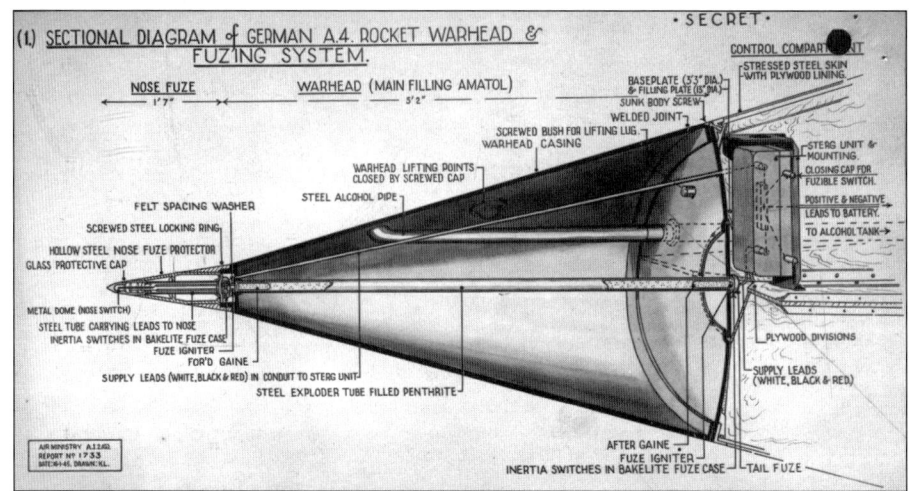

(*The National Archives: ref. AIR 2/9224*)

Unexploded V2s in England

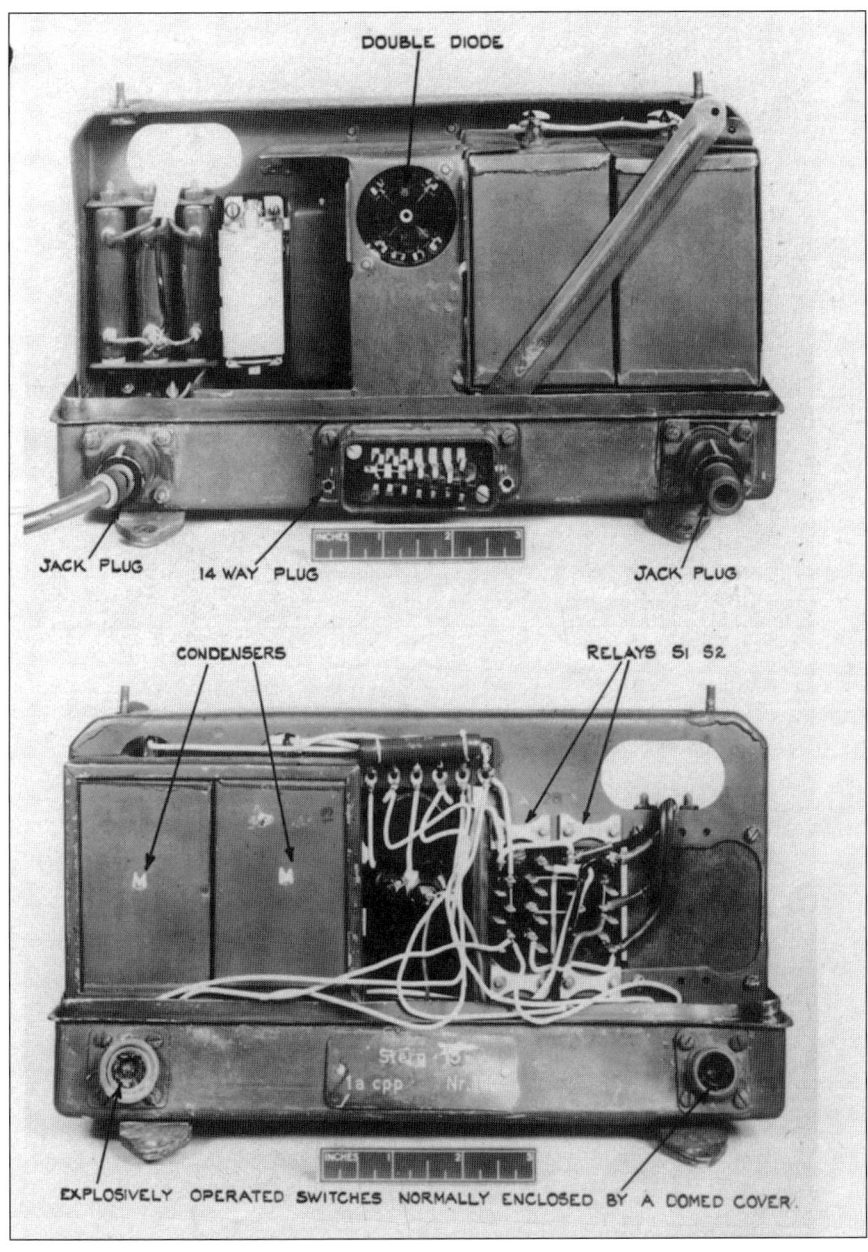

With unexploded V2 warheads it was essential for the bomb disposal officer to ascertain if the sterg unit (above) was still connected to the back of the warhead. The condensers in this unit could hold an electrical charge that would detonate the warhead should the fuzes be interfered with. It is interesting to note that the fuzing system in the V2 was designed in such a way that if the engine failed in the first 40 seconds then the fuzes should not detonate the warhead as it returned to earth. This safety feature was incorporated to protect the launch crews to some extent. Despite this, it was possible that a violent impact with the ground might still detonate the warhead without the fuzes operating. (*The National Archives: ref. AIR 2/9224*)

The nose fuze (right) and rear fuze (below) were pretty similar in design in that they both featured two trembler switches, set at right angles to each other. The nose fuze however, was also connected to an impact switch at the tip of the rocket. (*The National Archives: ref. AIR 2/9224*)

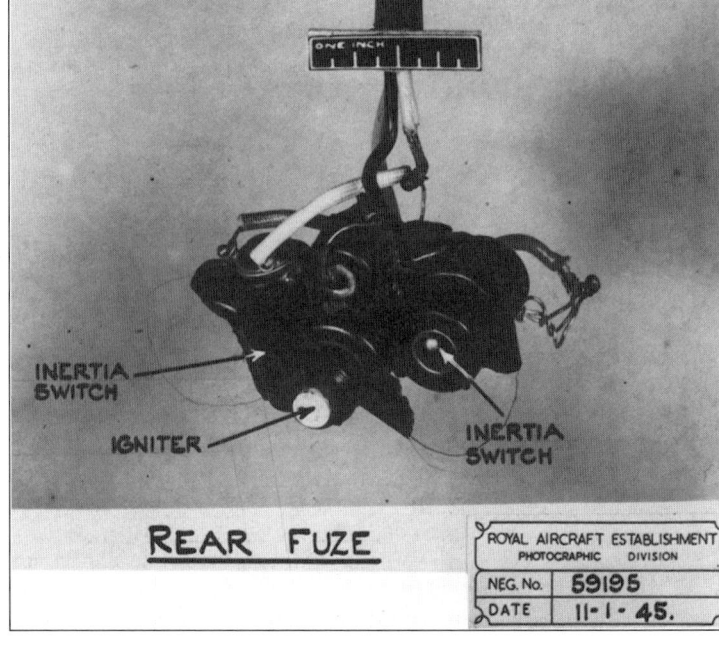

The sterg unit on this particular V2 was easily accessible. Major Gerhold was joined by Lieutenant Warner Charles Swinson, GM Royal Engineers.[17] The two men moved earth away from the crushed wreckage to expose the spaghetti of wires. They had to take great care as each wire had to be individually traced and kept clear of any other wires or metal surfaces, in case they had been damaged in the impact. A bare piece of wire could short-circuit causing the warhead to detonate. Once they had all been exposed and identified, the power leads from the sterg unit could be severed so no current could pass to the fuze(s). Again X-Rays were taken in order to ascertain whether there were any hidden booby-traps. Though they showed nothing of concern, as a precaution the detonating mechanisms were removed by remote control. With the exploder tube still in place in the warhead, the main filling of cast amatol explosive was steamed out.[18] This too was a delicate operation as the exploder tube was filled with a sensitive penthrite explosive. Next, a collar at the rear of the warhead was unscrewed, which allowed the removal of the booster charge from the exploder tube. The empty warhead and fittings were subsequently handed over to the Royal Aircraft Establishment for further examination.[19]

As the operation was being wound up, reports came in of yet another unexploded V2, again in Essex.[20] This one, again from the Statenkwartier area

Looking at 45 Northumberland Avenue, Hornchurch, in 2010, one would have no idea of the dramatic events that unfolded at this address during the Second World War, when a bomb disposal squad dealt with a V2 warhead embedded in the garden. (*Author collection*)

of The Hague, had broken up in the air at around 0540 hrs on 20 March. The warhead fell to earth in the garden of 45 Northumberland Avenue, Hornchurch.[21]

A ten-year-old schoolboy living in the house opposite came out in the morning to find the road covered in clods of earth. He was hoping for a souvenir, but had to go to school before being able to get a piece of the rocket and the area was soon cordoned off.

It was found that on hitting the ground the warhead had split open and the cast amatol explosive filling shattered. The local electricity supply company turned off supplies to the area to reduce the risk to the bomb disposal men tasked with clearing the site. Major Gerhold's men were able to remove the exposed explosive by hand and the sterg unit that was disconnected. This was sent off for evaluation and found to be still holding an electrical charge, so capable of detonating the fuze even after impact.[22] The warhead was removed without incident. The Paglesham V2, the first unexploded example, was the last to be removed. It was found 37ft down in the Essex soil and was dug out on 7 April.[23]

Two of the three V2s known to have fallen in the UK without exploding had broken apart while still in flight. These break-ups were not such an unusual occurrence, though often the warhead would explode in the air. The wreckage would nevertheless cause death and injury as it fell to earth and sometimes the warhead would detonate at a low altitude, its blast destroying property and lives on the ground. Such was the case in the London borough of Kensington, when on the evening of 12 December 1944, a V2 had an air-burst at a low level. In an ARP report of the time this was said to have been caused by the 'fall of a part of the rocket'. It left a crater at the north-east junction of Treadgold Street with Grenfell Road and another part of the rocket destroyed a two-storey terraced house on the north side of Treadgold Street. The blast from the warhead caused damage beyond repair to 20 houses in the area and resulted in the deaths of at least 2 people, injuring another 60.[24]

The British press were aware quite early on that some of the V2s suffered technical defects that meant they exploded in the air. The *Daily Mail* ran a headline, 'V2 Vanishes at 52,000 Feet Up'. The article went on: 'German efforts to get their rocket V2 into operation against this country appear to have met with some entirely unforeseen difficulties. There is very good reason to believe that a number of these projectiles have been launched, but that nothing more has been heard of them.'

The writer concluded by saying: 'Either the casing is not strong enough to withstand the lack of outside pressure at maximum height, or as it gathers speed on its downward flight the rocket creates such friction that the heat detonates the warhead. Indications are that to overcome this weakness the Germans have among other things reduced the size of the warhead from ten to six tons.'

Another example of a V2 breaking up in the air occurred on 28 October 1944, above Woolwich Common. The rocket apparently disintegrated, parts falling in Craigerne Road and St John's Park and at the eastern end of Vanbrugh Park, Blackheath, as well as the south-eastern garden area of Greenwich Park. Parts of it were taken to Lea Green Police Station. The ARP report intriguingly says that the warhead from this rocket 'finally detonated' in Pagnell Street, Deptford, by New Cross Station, which is 2 ½ miles away from where the rest of the missile came down. The warhead had actually fallen on open ground at the rear of houses at the southern end of the street and though it only made a slight indentation in the ground there, caused the demolition of about 12 two-storey houses and badly damaged many more. At least 5 people were killed as a result of this V2 and many others injured.[25]

The lift-off weight of a V2 was some 14 tons, though it rapidly lost weight as fuel was consumed, which had the effect of increasing the rocket's speed. Bear in mind the rocket could reach a height of 50–60 miles and an impact speed of three times the speed of sound! It was reported that a sonic boom, or crack, was sometimes heard just before impact and bizarrely the sound of the incoming rocket was then heard after the actual explosion, as the sound caught up.[26] Any air-burst or break-up in the air would still mean some very large chunks of rocket were going to hit the ground at great speed.

On 16 September 1944, a V2 air-burst over Southgate in North London and the disintegrating rocket fell over a wide area. The warhead exploded on the ground in front of houses on the north side of Tewkesbury Place. A

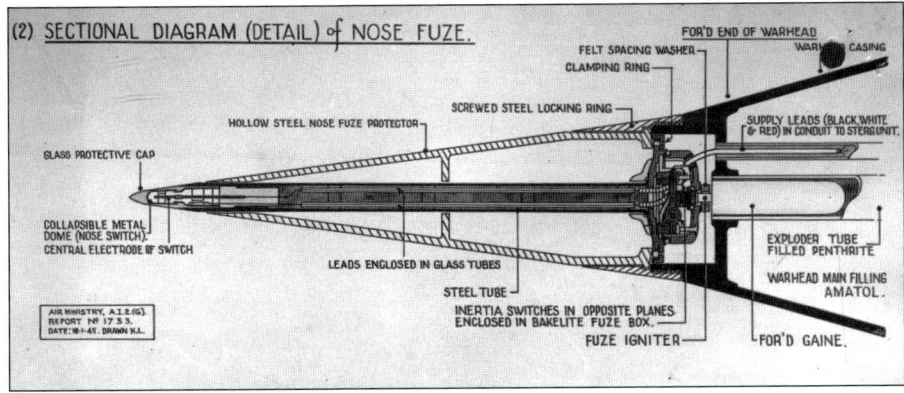

(*The National Archives: ref. AIR 2/9224*)

A relic from an air-burst in the author's collection. This electric motor is from a V2 that exploded in the air near Burnham-on-Crouch, Essex. It was part of the directional control system. One of these motors can be seen in the upright position at the base of the cut-away V2, bottom left of the photo. (*Author collection*)

large piece of rocket caused the collapse of the front part of a house at 98 Maidstone Road (though there is no visible evidence of this when looking at the house today), and an acid tank hit 128 Warwick Road. Various parts of the rocket were recovered from a mile and a half away. At least 12 people were killed by this rocket and a couple of days later the authorities still hadn't accounted for 4 others.[27]

The reader may find references in other books to a number of other unexploded V2s in the UK, but the evidence for these is somewhat lacking. For instance a V2 was reported as scoring a direct hit on the Southend Pier Pavilion, passing through the roof and the floor before embedding itself in the mud below without exploding. The rocket was later apparently salvaged and taken to a secret location for investigation. However, a photo of the unexploded V2 actually shows just the combustion chamber, that normally stayed pretty much intact after a detonation.[28]

It has also been said that an unexploded V2 was reported to have fallen at 1443 hrs on 7 December 1944, at Chediston, Suffolk – 2 miles west of Halesworth. The time doesn't tally though, with any known V2 impacts, and no information to back up the claim has been discovered. Likewise, a V2 warhead from a rocket that broke up in the air was reported to have fallen unexploded at 0247 hrs on 12 October 1944, at Dengie Marshes in Essex. No record has been discovered for this, but curiously the author came across a clump of fibreglass at an antiques fair a few years ago, with a note saying it

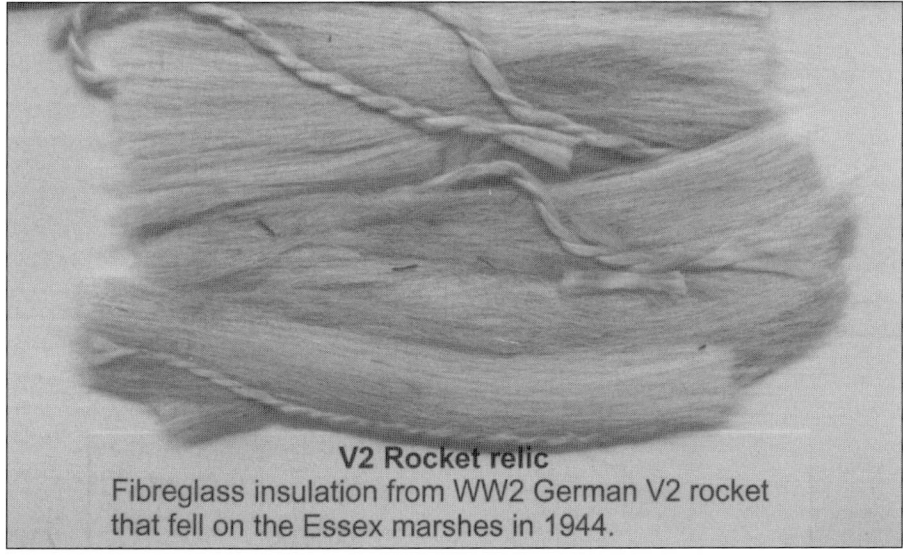

V2 Rocket relic
Fibreglass insulation from WW2 German V2 rocket that fell on the Essex marshes in 1944.

(*Author collection*)

was from a V2 rocket that fell in an Essex marsh in late 1944 and that it was recovered by Royal Navy Bomb Disposal experts in 1975!

In recent times, on 29 March 2012, the *Daily Mail* reported that an unexploded V2 warhead had been found 300ft from the shoreline, in the mud of Harwich harbour. The article dramatically described how a six-man team of bomb disposal experts were working urgently to make sure the rusting weapon's one-ton warhead did not detonate. The piece also provided an eye-witness account from Reuben Day, an 82-year-old man who remembered the V2 coming down in the coastal mud flats on the River Stour, between Felixstowe and Harwich. During the war he had been working as a fisherman in the area and recalled hearing an explosion overhead. He saw two clouds of black and white smoke and a white vapour trail. Another fisherman in a boat was apparently swamped as the wreckage of the missile fell into the water close to him. The next day a local policeman was taken out to the spot by Mr Day, but that was the last he knew of the matter.

The piece of V2 wreckage continued to protrude from the mud for the next few decades and was well known to local sailors. Then in March 2012 the 'obstruction' was discussed at a meeting of harbour users. As a result it made the news and the Royal Navy's Southern Diving Unit 2, commanded by Dan Herridge, were called on to investigate. Army BD personnel from 101 Engineering Regiment were also involved, providing additional excavation equipment.

The wreckage was only exposed for about two hours at low tide. However, the Navy quickly identified the rusty remains as the combustion chamber from a V2 rocket. They then set about uncovering it further, to discover whether or not any explosive was present. However, within a couple of days it was apparent that the wreckage consisted only of the engine. It was subsequently hauled out of the mud and was to be donated to the local sailing club.

This rusty relic probably originated from a V2 launched at 1557 hrs on 10 October 1944, from Battery 444 at Rijsterbos, Middenleane in the Netherlands, as that particular missile is recorded as having exploded 1,200 meters above Harwich harbour.[29]

Another V2 was said to have broken up over Tollesbury, Essex, on 15 March 1945 and produced no blast crater. It was assumed by the authorities that the warhead had fallen in the River Blackwater, but that was never proven.[30]

The final V2 to fall in the UK came down at Court Road, Orpington, on 27 March 1945. The explosion killed one person, Mrs Ivy Millichamp. She was in fact the 60,595th and last civilian to die in the UK in the Second World War as a direct result of enemy action.[31] It is estimated that some 2,754 civilians were killed in London by the V2s.

Unexploded V2s in England 85

As a teenager in the war Reuben Day witnessed a V2 break up over Harwich Harbour. In 2012, a Royal Navy bomb disposal unit investigated wreckage from that missile, but found no explosives. All that was discovered was the V2's engine. (*John Cox*)

The engine was recovered with the help of the Royal Engineers. (*John Cox*)

Despite the fact that from the end of March 1945 there were no more V-weapon attacks against the UK, on the Continent they continued to cause death and destruction for some time to come.

Chapter 4

Europe

Bomb disposal units were among the forces that landed with the Normandy invasion. Although they were not part of the first attacking wave, one BD man, Corporal Middleton of 6205 BD Flight, is known to have died on 6 June 1944 and is buried at Bayeux War Cemetery.[1] On that day at least one unexploded bomb is known to have been disposed of. It had hit one of the LSTs (Landing Ships, Tank) and gone through the hull and two bulkheads without exploding, killing one man in the process. It was successfully disposed of by members of the army on board.

The BD units themselves were to land close behind the main attacking force as their expertise was required to ensure that areas captured, such as ports or airfields, could be promptly cleared of booby-traps. This was essential if the lead forces were to be quickly resupplied and reinforced. A case in point would be the Royal Engineers BD unit that was transported across the Channel on 6 June on HMS *Eagle*. *Eagle* arrived late in the day and while off the French coast that evening was lucky to survive an air-raid in which the ship next to her was hit. After a delay due to rough seas the men finally made it ashore when the ship moved on to Port-en-Bessin, a small fishing harbour west of Arromanches. Their first job was to assess two enemy boats in the harbour, one of which was on fire. They had to check if there was ammunition that was likely to detonate, including booby-traps or time-delayed bombs. The surrounding buildings were also checked.[2]

Other landings for the BD personnel were not without incident. Another Royal Engineers BD section ran into trouble as they drove off a landing craft on 10 June – the water was too deep and their vehicle was 'drowned'. The men were picked up by the Royal Marines, placed on a wreck and told to wait until low tide, some seven hours later.[3]

Some RAF BD personnel were to suffer a much worse fate. On 7 June, Flight Lieutenant Cartwright and 6225 BD Flight headed towards the French coast. In the early hours of the 8th their convoy was attacked by E-boats and shelled by coastal batteries positioned at Le Havre, 4 miles away. A shell hit the LCT and it sank in less than two minutes. Another landing craft, believed to be LCT1025, rescued some survivors and set them ashore at MIKE/NAN beaches. The BD unit's war diary states that seven of their personnel were missing, six were hospitalized and one was taken prisoner of war.[4] LCT 390 had gone down with 90 per cent of the unit's equipment, including a number of vehicles.

Once they had made it onto foreign soil the British bomb disposal units were able to get to grips with the enemy. The nature of their duties meant that they often worked in areas where battles had already taken place, but there were times when they would be under fire and on occasion they would come literally face to face with the enemy. The first instance of a Royal Engineers BD unit taking German prisoners is believed to have been on 16 August near Caen, when Lieutenant E. J. Beer of 24 BD Company captured two enemy soldiers.[5]

In the period just after D-Day the units tended to concentrate their efforts on clearing mines and booby-traps so that the Allied forces could move more freely. After a while, other munitions would feature more prominently in their work. As the Allies moved on, so they came across the V1s and V2s. Some were found at their launch sites or still in storage. Others had been abandoned while in the process of being moved or at their assembly plants. Some had been fired and had come down unexploded.

Most of the BD officers involved had only ever seen the V1s as they passed overhead but now they would be forced to confront unexploded examples on the ground. It was not long before the war diaries of various units reported their first dealings with these missiles.

On 4 September, 224 BD Section's records state that they removed the remains of two crashed flying bombs from an airstrip at Realcamp, about 24 miles from Dieppe. Then on 7 September another unexploded V1 was dealt with by the section at Delettes. Bomb Disposal Officer Lieutenant Boorman found that this bomb was fitted with a 106* fuze and two other fuzes. One of them was a clockwork fuze that started ticking. It was decided at this point to blow the V1 up in situ.[6] This was almost certainly Lieutenant Boorman's first experience of attempting to disarm a V1, as he carefully made a drawing of one of the fuzes. After he had blown up this bomb his unit had to spend the rest of the day repairing the roofs of nearby buildings damaged in the explosion.

When the Germans retreated they destroyed stockpiles of V-weapons rather than let them fall into Allied hands. Major C. R. Wood of No. 23 BD Company visited a flying bomb transit dump on 11 September 1944, that had been blown up by the Germans near Rennes, France. The demolitions had been hastily prepared and had not destroyed the dump completely. At the site he found four unexploded warheads minus fuzes, one nearly complete V1

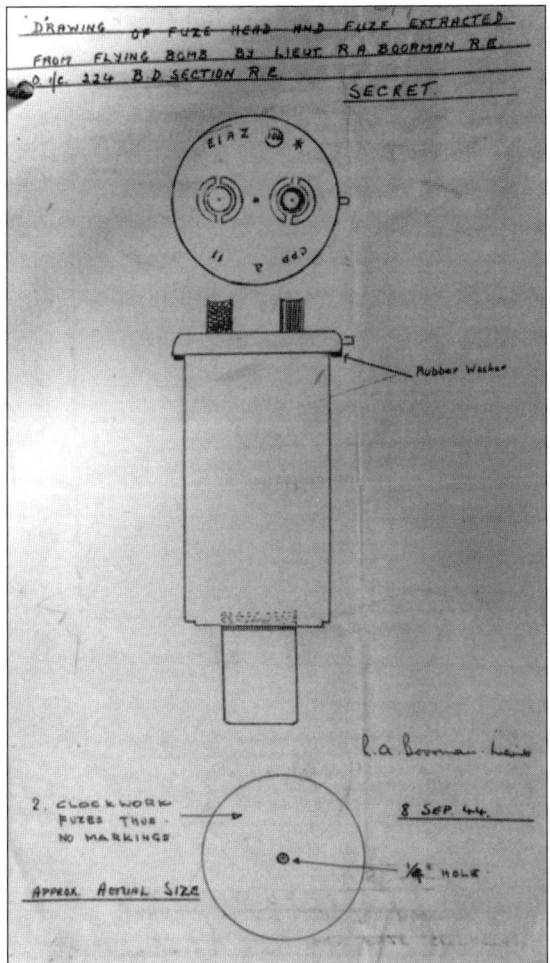

Drawing of fuze head and fuze made by Lieutenant Boorman. (*The National Archives: ref. WO 171/1977*)

with no nose cowl or fuzes and two that were damaged, fitted with 80A fuzes. The J process was used to neutralize these. A couple of days later he was asked to look at more V1s that were loaded on a train at Schaffen. Fortunately these were all in a relatively safe condition and could be removed without problem.[7] Major Wood had been working in BD for some time and had already been awarded the George Medal. Assisted by Sapper J. Williams he had recovered one of the first clockwork time-delay fuzes from a bomb at an electricity depot in August 1940. A month later both Major Wood and Sapper Williams were injured while dealing with another unexploded bomb.

On 27 November 1944, Lieutenant Beswick of 53 BD Platoon disarmed a crashed V1 at map reference L044298, near Rouen. It was fitted with two 80A fuzes in the side pockets and a 106 in the central one.[8] It is thought that this

V1 wreckage in Mehren, Germany. (*Alois Mayer*)

bomb was removed by a section from 59 BD Platoon, also based at Rouen, a couple of days later.

Another unexploded V1 was discovered near Thielt, Belgium, just before Christmas. A reconnaissance was made by Major Bainbridge of 25 BD Company, along with Lieutenant Curry, on 22 December. The V1 was disposed of the following day.[9] Major Bainbridge was another of those who had distinguished themselves in bomb disposal, had been awarded the George Medal and had spent time in hospital. His injuries were caused by an explosion that occurred while he was removing a clockwork time-delay fuze from a large bomb near a main-line railway station at Woolwich, London, four years earlier.

The Americans also started to come across unexploded V1s. Lieutenant Leo E. McCollum of the US 951st Battalion recalled that on 2 September 1944, while positioned north-east of Jeantes, near Plomion, a number of V1s flew overhead. As one was going over its engine cut out and everyone dived for cover. Though it crashed into the Battalion's area, it did not explode. As the men stood up to have a look they noticed that Major Boyd Clark's trousers were round his ankles. He had been on the other side of a hedgerow answering a call of nature when the V1's engine stopped. The hedge was about 4ft high

Two photos of a V1 being examined by American troops. It came down near Jerly Command Post, Plomion, France, on 2 September 1944. (*US Army Military History Institute, Carlisle Barracks, Pennsylvania / NARA*)

and very dense at the top – Major Clark had no idea how he was able to clear the hedge with his trousers down. Everyone had a much-needed laugh, including the Major. Lieutenant Russell Kelch reported the unexploded V1 and within a couple of hours a British bomb disposal squad arrived to defuze it.[10]

This V1 came down without exploding at Grimbergen, Belgium some time between January and April 1945. (*Author collection*)

The RAF's 302 (Polish) Squadron had a similar experience. While based at Grimbergen in Belgium, the men heard a loud swishing noise in the night. They went to investigate but could not see anything in the darkness.[11] The following morning they found an unexploded V1 and beat a hasty retreat until it was defuzed by an RAF bomb disposal team. This could well be the same missile that Polish war correspondent Wladyslaw Kisielewski heard hitting the earth with a thud in the middle of the night. He reported that its wings broke off and the fuselage came to rest by the doorstep of his quarters. He went on to say that men found themselves climbing over it in the darkness and what a shock they had as they realized what it was![12]

The failure of V1s to reach their target, crashing shortly after take off, appears to have been a relatively common occurrence. As the Germans were forced to retreat so they had to look for new launch sites. The Eifel region was within range of Antwerp and the Germans, knowing how many V1s crashed on launch, felt that German civilians were less likely to be affected by launch failures in this sparsely populated region. Residents in the area soon became accustomed to the spluttering sound of a malfunctioning V1. Launches were made towards the cities of Brussels, Antwerp, Liège and Amsterdam. Some believed that 30 per cent of the V1s fired actually crashed within a 60km radius of the launch site and locals began to fear the 'revenge weapon' more than the

The V1 shown in these photos had only managed to travel a short distance from its launch site at Foucarmont (between Dieppe and Amiens). It is here being inspected by a Canadian sergeant and a member of the French Resistance. This particular V1 was seen by the British public at the cinema in a Pathe Gazette newsreel entitled 'V1 Sites Captured'. The film was released in early September 1944 and it was pointed out that 'a considerable number of flying bombs crashed before they were properly airborne'. (*Author collection*)

Allied bombers.[13] Their fear was not without foundation. On 4 November 1944, one crashed soon after takeoff and landed on the village of Gess, destroying an entire street and killing 12 people.[14] A couple of weeks later on the 27th, a number of Hitler Youth children from Hanau were killed when another came down on the road between Hillesheim and Oberbettingen.[15] Yet another incident occurred on 14 January 1945, at Junkerath. This V1 skidded along the ground and was deflected by a chestnut tree into a house, killing the occupants – a mother and son.[16]

One of the reasons for so many crashes at this time was the 'improved' autopilot, which was designed to allow the missiles to be fired in a direction other than at the target. Once in the air a single course alteration should have turned the V1 towards its target, but this sometimes had disastrous consequences as the V1 lost control.

A malfunctioning V1 was seen to circle the Flemish village of De Klijte on 27 July 1944. After 30 minutes it hit the ground without exploding, landing in a potato field. An unfortunate farm employee, Camiel Dequeker, was struck by its wing and killed. Four days later the Germans blew it up, causing damage to the nearby farm buildings in the process.[17]

Another example of a V1 blown up in situ occurred on 15 November 1944, at Heemstede in the Netherlands. This one had come to rest close to St Bavo Church, in a garden facing the residence of the Roman Catholic priest. Instead of defusing it the Germans chose to blow it up, after evacuating people from the nearby houses. Many of their homes were destroyed in the blast and much damage was done to the church. The stained glass windows on the eastern and southern sides were destroyed.[18] The church has since been restored though not all the stained glass was replaced.

According to locals living near launching sites the Germans sometimes salvaged the missiles that could still provide some useful spares and blew up the remainder. This fact was evident too from documents captured by the British after the Normandy landings, which included instructions relating to failed V1s. They stated that any V1 that had come down unexploded close to the front line and likely to be captured intact, was to be destroyed with minimum delay, and care should be taken to check the surrounding area to make sure that, if it had broken up on landing, all parts were found and destroyed. It was also noted that jarring and pressure on the warhead should be avoided and a specialist capable of disarming it should ideally be called for. The instructions went on to say that before blowing up the warhead an attempt should be made to ascertain the white six-digit number painted on the back end of the explosive container.[19] From this the Germans could trace if there was a manufacturing problem such as sabotage or a bad batch of components, etc.

The launch crews of both V1s and V2s were concerned at these failed launches, fearing the tell-tale scars on the landscape might give away their positions to enemy aircraft. Accidents did occur quite regularly at launching. Some rockets barely got into the air at all before mechanical troubles manifested themselves. One V2 lost power and simply toppled over. The launch crew are said to have used pickaxes to puncture the rocket in order to drain the fuel and, rather unsurprisingly, the fuel ignited and exploded, killing three and injuring others.[20] Whether this is true or just folklore is open to debate.[21]

The fear of being spotted by Allied aircraft meant the Germans often launched their V2s from small woodland clearings. Care had to be taken to ensure the clearing was actually large enough. There was one case where the fin of a V2 clipped a tree on take-off, causing the rocket to go out of control. It went on to crash 7km from the launch site, and on this occasion the warhead failed to explode.[22]

On 30 September 1944 a V2 fired from a woodland site at Murnserleane in Holland rose some 600ft, then suffered a mechanical failure that resulted in it falling back to earth. A number of the firing crew were injured as the fuel tanks exploded. The warhead, engulfed in flames, exploded 45 minutes later. A small shrine known as 'the little peace temple' was destroyed by this particular rocket.[23] There were apparently a couple of other V2s launched from the same site that failed, also coming down in the surrounding woodland.

News of these failures in Holland made it into the British press, and articles appeared such as the one in the *Sunday Pictorial* newspaper of 12 November 1944, headlined 'V2 Crews Killed as They Fire'. Another headline in the *Daily Telegraph* of 29 November read 'Boomerang V2 – Fell Back on Site and Exploded'. This article described a 'German rocket bomb' launched from a point north-east of Arnhem as partially exploding in the air and then falling back on the launching point, where it exploded violently again. It commented that this 'boomerang' V2 was reminiscent of some of the V1s which had made return trips to their launching sites, as witnessed by two Canadians, Flying Officers Robert Fullerton and Peter Castellan, while on a night patrol from their 'Night Hawk' Squadron base in France.

Residents in the Zwolle launching area in the Netherlands made a note of at least 15 'misfires' in January 1945. These included V2s that only went a short distance and ones that failed at launch. Six detonated at the site between ten minutes and an hour after the attempted launch had taken place. One was said to have 'burnt as it stood'.[24]

The Allies discovered evidence of these V2 failures when they overran the launch areas. At a firing position discovered in woods on the estate of Baron Von Vorst, at Mataram in the Netherlands, they found a V2 had failed or fallen at the entrance to a fire control vehicle shelter. It had caused a crater that had half filled with water. In and around the crater were found, besides part of a V2, several fragments of a vehicle which may have been the fire control vehicle itself.[25] No wonder the morale of the launch crews suffered.

At Hellendorn the Allies received several reports of 'prematures', and it appeared that many rockets could be heard to strike the ground with a dull thud and that no explosion took place for at least five minutes, thus giving everyone in the neighbourhood time to open all their windows. Witnesses varied in their estimate of the time-lapse generally between the rocket falling and the explosion, usually 5 to 15 minutes, but all exploded within 20 minutes except one, which for reasons unknown did not explode until the following day.[26]

Sometimes a misfire could be quite spectacular. One V2 was seen to tip over on launch and then fly horizontally at an altitude of about 14 metres. The rocket continued like this until it crashed some 6km away, near the town of Hachenburg, Germany. The warhead did not explode but was destroyed later in a controlled explosion.[27]

It was not always plain sailing either for the German bomb disposal officers called in to disarm V-weapon failures. Lieutenant Bube was a Luftwaffe officer serving in the Cologne district. He was an experienced man and had dealt with about 800 Allied bombs, but in actual fact was not too familiar with some of the German ordnance. One day he was sent to Schelden to deal with an unexploded bomb. When he arrived he found a crashed V1 that had broken up, its explosive lying nearby. He removed a No. 80A fuze, but as he unscrewed the small explosive gaine from the end of it, the fuze detonated. A piece of it penetrated his head, killing him instantly.[28]

As the Allies pushed on they discovered V-weapon production and assembly facilities. US Army Captain John Feldman (mentioned in Chapter 1 at the Strawberry Hill Farm V1 crash site) wrote a report to Colonel Kane on 23 September 1944, in which he described the V-weapon assembly plant at Nucourt, near Magny-en-Vexin (30 miles north-west of Paris). In it he states that the area, a one-time quarry, had extensive damage, having been heavily

Captain Gordon A. Ruesink was in command of the 81st BD Squad. This unit was stationed at Asch, Keerbergen and Brasschaat, Belgium, during February through March 1945. They made a name for themselves at the Ninth Air Force HQ for the number of V1s and V2s they disposed of. (*T. Dennis Reece*)

Europe 97

These photos show a couple of V1s that came down at unidentified locations on the Continent. (*T. Dennis Reece*)

bombed by the USAAF (US UXBs were present) and that it was impossible to examine more than about one tenth of the installation. In one room there were apparently six undamaged warheads in their crates.[29]

The American 9th Air Forces operating in mainland Europe were supported by their own bomb disposal squads. They were split up to cover three different

Captain Thomas E. Reece of 10th BD Squad (left), with Lieutenant George W. Collins of 77th BD Squad at St Trond, Belgium. The two were good friends but Collins was killed on 16 February 1945, just three days after this photo was taken, while he was working on a German Riegel 43 mine. Reece landed in Normandy in August 1944 and regularly visited squads in his sector who were working on V1s. From October to November 1944 he was stationed at Liège, then went to St Trond, where he remained until the end of March 1945. (*T. Dennis Reece*)

Europe 99

A crashed V1 that has flipped upside-down – notice the fixture for hooking it on to the launching piston. (*T. Dennis Reece*)

Thought to be Technical Sergeant Thomas J. Roache of the 73rd BD Squad holding an 80A All-ways fuze. This unit was stationed at Florennes-Juzaine, Belgium, from mid-September 1944 to late March 1945. (*T. Dennis Reece*)

Captain Reece and Captain E. H. Jamieson III, CO of the 73rd BD Squad standing in front of the wreckage of a crashed V1. (*T. Dennis Reece*)

A disarmed V1 warhead – notice the two fuzes and the picric acid booster charges (white drum-shaped objects) next to them. (*T. Dennis Reece*)

areas. Unexploded V-weapons had to be dealt with as they were found. For example, in November the 74th BD Squad based at Juvencourt in France, came across and disarmed an intact V1, which was then dismantled and sent by air to the UK for RAF experts to study.[30]

On 31 January 1945, the 787th Automatic Weapons Battalion shot down a V1 that failed to explode. Captain Wally Collins, then stationed at Keerbergen to the south-east of Antwerp, was called in to dispose of it. He found that the V1 had been fitted with a cardboard tube containing a roll of propaganda leaflets that were to be blown out by a small charge of black powder. Though this one was successfully dealt with, Collins was not lucky enough to survive the war. He was killed a month later while working on a mine.[31]

The first recorded recovery of an unexploded V2 by the Allied forces occurred at the end of September 1944, when on the 27th an unexploded warhead was found in Belgium in the US First Army's area. 1st Lieutenant S. F. Rausch of Ordnance Technical Intelligence Unit E and Doctor L. F. Woodruff, a technical advisor, were given the job of examining it the next day. They found the warhead almost completely buried. It had actually gone down 10ft and along about 20ft before coming back up close to the surface,

The first unexploded V2 warhead with its casing peeled back like a banana. (*The National Archives: ref. AIR 2/9224*)

Wreckage from a V2 collected from farmland at Spa, Belgium, some of which was sent back to the UK to be further examined. (*The National Archives: ref. AVIA 6/25652*)

leaving a trail of powdered explosive behind it. The warhead was dug out and sent by truck to Com. Z. Ordnance Depot near Paris. Samples of the explosive and booster materials were sent to Picatinny Arsenal, USA, and RAE Farnborough for analysis, though the greater part of it was removed and destroyed.[32]

A month later, on 29 September, a number of large components from a V2 were reported to the US 12th Army having been found in the vicinity of Spa in Belgium. Experts from RAE Farnborough, accompanied by a US Army major, visited several farms in the area to examine the wreckage. The local French Resistance assisted, as they were aware of the various locations where the debris had come to rest. The actual crater was extremely small, measuring only 10ft wide by 6ft deep and at first it was thought that the warhead had not actually exploded. However, a piece of the warhead casing was later found embedded in a tree some distance away. From the parts discovered it appeared that the alcohol tank burst in the air, causing the warhead to separate from the rest of the rocket. Many of the parts were sent back to Farnborough as they were in better condition than material previously recovered.[33]

On 19 December 1944, a second unexploded V2 warhead was found in much better condition. Squadron Leader Kenneth Scamell, of 5139 BD Squadron, along with Flight Lieutenant Fleming and Flight Sergeant Martin, Corporal Oram and a party of airmen were sent to Lierre, Belgium, to recover it.[34] They set off early and were followed by Corporal Hauxwell and Corporal Briggs with a steam generator to enable the squad to steam out the explosives.

It was assumed that the rocket had broken up at altitude and the warhead had come down without exploding. It had impacted sideways, passing through and scattering a store of root crops, and had come to rest in a small crater just beneath the surface some 20 yards from a farmhouse (thought to be in the vicinity of Groenedaellaan). On digging down it was found that the rear of the warhead casing had split open and spread.[35]

Once the missile had been partially excavated the main filling was removed by hand. Though the forward part of the missile was in good condition, the removal of the nose fuze presented some difficulties. It was tightly secured by a heavy tapered locking ring and grub screw. The grub screw came out all right, but the locking ring was stuck fast. The threads were treated with penetrating oil which was allowed to soak in overnight, while a local blacksmith constructed a 'C' spanner with an extension tube nearly 3ft long in order to get some leverage.[36]

The following morning, 20 December, penetrating oil was again applied to the threads and the 'C' spanner fitted. After considerable effort on the part of Squadron Leader Scamell, the locking ring unscrewed and slipped over the truncated cone-shaped fuze. It was then seen that the fuze, which had been firmly held by hand throughout the proceedings to prevent movement, abutted a 4in. long explosive gaine and was attached by only two of the three leads described in an Air Ministry instruction booklet. It was thought at the time that this third lead had never been connected and was probably the reason it failed to detonate. Separation of the fuze was then easily effected. Scamell

Squadron Leader Kenneth Scamell of 5139 BD Squadron worked on one of the first reported unexploded V2s. (*RAF Bomb Disposal Association*)

The recovery of this V2 warhead was done by members of 6206 and 6210 BD Flights. Some of the main explosive filling was removed by hand (picture shows one man's head and shoulders are inside the warhead). (*The National Archives: ref. AIR 2/9224*)

made a partial examination of the fuze to ensure it could be certified safe for transportation back to the UK.[37]

The warhead was then removed from the excavation and cleaned up by means of a steam generator. This was also used to steam the explosives out from the exploder pocket. All parts of the warhead and a large sample of the explosive filling (thought to be amatol, from the external markings on the casing) were sent back to the Air Ministry the next day for further examination. They were accompanied on their journey by a technical intelligence officer with instructions to deliver it personally.[38]

The Royal Aircraft Establishment at Farnborough and Air Intelligence branch A.I.2(g) both issued their own reports of their findings in January 1945.[39] The firing system was acknowledged to have been so efficiently conceived that these two warheads were the only ones to have failed since the V2 attacks began on 8 September 1944 – that's just two known failures out of approximately 1,150 V2 incidents up until that date.

The explosives from these warheads were analysed and found to be cast amatol (60 per cent TNT and 40 per cent ammonium nitrate). This was actually one of the least powerful blast fillings. If aluminized fillings were used, the blast force would be increased considerably. It was assumed by Allied intelligence that the Germans used amatol because it was insensitive to the heat generated by the rocket in flight. The investigation team at RAE

Farnborough were very interested in the effects of temperature on the warhead, though they realized the samples they had were from defective V2s which would not have been travelling at the same velocity as those that performed normally.[40]

The Farnborough report also looked at its vulnerability to shell splinters and concluded that damage to the main structure of the rocket by shell fragments might cause it to break up in the air, but would be unlikely in itself to cause a premature detonation of the warhead.

A third report also written by RAE Farnborough at the time focused just on the fuzing system and how it functioned.[41] It explained that there were two similar fuzes, one at each end of a central exploder tube. Each had two inertia switches and the front fuze was also connected to a contact switch at the extreme nose of the rocket. The electricity to fire the fuzes came from condensers in a 'sterg' unit (Sterg was written in white on the side of it). This unit was located immediately behind the warhead and its electrical supply came from the rocket's 32V battery. The sterg unit was an important feature to the bomb disposal men. If it was still attached to an unexploded warhead after impact, then it was still capable of firing the fuzes, because it could hold an electrical charge even when the leads to the main battery had been severed. To safely disarm the warhead the BD officer would have to remove the jack-plugs from the sterg unit to isolate the fuzes completely, or ensure that three wires were cut and their ends insulated. The first one to be cut would be a white wire, followed by a black one – these were leads that fed the DC electrical supply to the fuze. Then a red wire was to be cut. This was connected to the inertia switches and allowed for a proximity fuze to be wired into the circuit.

The report acknowledged the fact that there were quite a number of V2s that broke up in the air and this appears to have been anticipated by the Germans to some extent and allowed for in their fuzing system. For example the sterg unit was mounted as close to the warhead as possible, so that there was a reasonable chance of it remaining attached if a break-up occurred (the condensers within held the electricity to fire the fuzes). It went on to say that airbursts may well have been due to the fact that as a rocket broke up the warhead could swerve violently. This would result in the inertia switches being activated, detonating the warhead in the air. It was thought that warheads where the sterg unit had detached during a break-up could still explode on impact. Though the power supply to the fuzes was lost, the velocity of the warhead as it hit the ground might be enough to cause a partial, or low-order detonation. Air Intelligence were aware of this, having seen evidence of it at a small number of incidents in open land on the Continent.

On Christmas Eve 1944, just a few days after working on the V2 at Lierre, Squadron Leader Scamell was involved in the recovery of another unexploded warhead, this time from a V1. It had come down in a rural area of Moerbeke, Belgium (roughly between Ghent and Antwerp). The warhead was practically undamaged though the nose fairing had become detached. On examination it was found that the electric 'belly' switch was missing (probably torn off in the crash landing). The nose fuze pocket had never had an ElAZ fuze fitted, though a number of picric acid exploder pellets were present, beneath a tightly rolled corrugated cardboard packing insert. This was held in the pocket by a wax-impregnated cardboard disc and locking ring. It was thought at the time that the lack of a fuze was an oversight by the responsible German armourer. The associated Ent 106 could not be found but this may have also been torn off in the landing. Both the side fuze pockets were fitted with 80A fuzes. As by now it was getting dark, only the rear fuze was treated to the 'J' process, to 'jam up the works'. The following morning the other fuze was also treated. They were then both withdrawn remotely using a long length of cord, just in case of booby-traps. It was not possible to perform a jerk-test on the bomb as the location was too difficult to reach by mechanized transport. In fact the V1's warhead had to be dragged by a team of horses to the nearest road about a mile and a half away! A later examination of the fuzes showed that both strikers had impinged upon the detonator caps but had not pierced them. The strikers had returned to their original positions, held by creep springs, and were now filled with the hardened plastic that had been injected during the 'J' process. As a result of analysing these fuzes, Squadron Leader Scamell produced a collet that would enable air pressure to be injected into the fuze after the 'J' process, so that the resin-hardener gel would penetrate every nook and cranny. This was sent to the Air Ministry for their consideration, as Scamell had found that in one of the fuzes the resin hadn't in fact reached the flash channel and gaine cavity. The warhead, with others, was later despatched to the UK for static detonation trials.[42]

The V1s were to keep 6210 BD flight busy over the following months. Just a few days after the Moerbeke V1, on 1 January 1945, they were 50 miles away at Diksmuide (10 miles south of Ostend), working on another. Two 80A fuzes were immunized by the 'J' process and removed by remote control by Sergeant Empett. The next day the warhead was taken to Sint-Denijs-Westrem aerodrome, Ghent, prior to being shipped to the UK.[43]

On 3 January 1945, a party from 6206 BD Flight RAF, led by Flight Sergeant Martin, left their base to check out an unexploded V1 approximately 20 miles west of Brussels at Ath. On arrival they found the V1 was completely wrecked – the site was cleared up and the explosives from it removed to a safe area.⁴⁴

Then on the 13th another was reported at Munkzwalm, 8 miles south of Ghent. The following morning Scamell, along with Flight Lieutenant Cox, Sergeant Empett and a party of other ranks headed for Munkzwalm, where they found a V1 that had come down unexploded, described as in 'perfect condition'. Again the fuzes were neutralized using the 'J' process – introducing a jamming-up liquid. The two 80A fuzes and a 106 nose fuze were removed and the warhead placed in a bomb dump to await shipment to the UK.⁴⁵ It was on the 21st of January that three V1 warheads left for the UK accompanied by Corporal Connolly of 6210 BD Flight (Connolly incidentally had had a close shave in the December, when the mine detector he was using caught the prongs of an S-mine in long grass. The mine fired into the air smashing through the pad of his detector but fortunately failing to explode in the air).

This V1 complete with warhead was discovered just north of Dahnen, Germany, on the embankment of a railway. Photographed by American forces on 22 February 1945. (*John Glascock, www.jajg.com*)

A snapshot in the author's collection of a V1 in its final moments before impact. This is thought to have been taken in the Liège area. (*Author collection*)

This V1 crashed shortly after being launched from a site at Fontaine-sous-Preaux, near Rouen, France. (*Author collection*)

Any facilities identified as being connected to the V-weapons were potential targets for bombing. This meant BD personnel had not only to deal with abandoned V-weapons in the area, but also unexploded bombs left over from Allied raids. An unidentified V1 ramp smashed by bombing (above) and the V1 storage and launch site at Siracourt, France (below), give some indication of the intensity of some of these bombing raids. Siracourt never actually became operational and some have suggested that it was really just a decoy site. (*Author collection, The National Archives: ref. AIR 40/2524*)

Not all the unexploded V-weapons reports were accurate and sometimes the BD organizations found on arrival at the site that the weapon had in fact exploded, as was the case at Termonde on 15 January 1945. Here Flight Lieutenant Fleming led a party to investigate a V1. It had been a direct hit on a farmhouse, killing four people. In case they had gone to the wrong location, Fleming contacted the Commissioner of Police and he confirmed that this was the only V1 to have come down in the area at that time.[46] Another exploded V1 was found on 26 January by Sergeant Ritchie and his squad at Hyzinghen. They did retrieve a number of leaflets that had been carried on this missile.[47]

A genuine unexploded flying bomb was found in the Ghent area on 31 January. Again, Squadron Leader Scamell led a team that included Flight Lieutenants Cox and Booth, Corporals Light, Gregory, and Williamson, Leading Aircraftman Stan Rubery and one other serviceman (possibly Corporal Hallows). Here they found a new type of V1, but the records do not go into any detail. However, it could well be the one referred to in a Bomb Disposal Intelligence bulletin issued to the US Army in February 1945 that stated, 'New type V1 encountered with plywood warhead containing three side fuze pockets and no fuze pocket in nose of warhead. In specimen examined forward fuze pocket contained no fuze but was covered by a wooden plug. New side fuze pocket may be either a new location for 106 star fuze or another fuze, possibly new fuze.' The bomb Scamell's team dealt with was fitted with two 80A fuzes. These were immunized and the warhead taken to their bomb dump pending shipment to the UK.[48]

The next reported V1 call-out for 6206 BD Flight occurred on 9 February 1945. Flight Lieutenant Booth phoned the unit to tell them a V1 had come down on a farm just east of Antwerp. This city was on the receiving end of a great number of V-weapons (over 4,200 V1s reached the Antwerp region, though many were shot down, and 1,600 V2s are reported to have fallen on the city).[49] For this particular V1 the instructions were that no fuzes were to be immunized – the unit should just take particulars and then report to the Squadron HQ. Flight Lieutenant Fleming, Corporal Melville and Leading Aircraftman Danks were tasked with this particular job. But on arriving at the crash site there was no sign of a crater. All they could find were small pieces of V1 widely scattered. The wife of the farmer told them that the army had already collected the warhead. It is thought that the missile had been hit by shellfire.[50]

One eye-witness living in a suburb of Antwerp records how at around 11.00 am on 9 February 1945 a V1 fell in the back yard of a house in Hoboken, in the south of the city. There was a loud bang but it did not explode. Neighbours said later that the street was cordoned off and US engineers from the Kiel barracks came with a big truck to load up the almost intact missile.

A few days later a message was received by 6206 BD Flight from 5139 BD Squadron that another two V1s had come down unexploded in the Antwerp district. Flight Sergeant Martin checked them out but found nothing of interest; both were completely smashed up.[51]

On 16 February 1945, Flight Lieutenant Booth and Warrant Officer Dyer went to Doel, just north of Antwerp, to look at a flying bomb that had crashed on land occupied by 997 Balloon Squadron.[52] It had come down close to the squadron cookhouse. They found that the warhead had partly broken up on impact and pieces of the plywood casing and quantities of explosive filling were lying around the crash site.[53] The following day Flight Sergeant Martin of 6206 BD Flight was told to take a party to deal with this, but on arrival he was informed that the Army was going to take care of it. However, by 3 March the Army had not done so and Martin was again sent to remove it. Two corporals and two leading aircraftmen accompanied Flight Sergeant Martin. By 2030 hrs the next evening the men returned from Doel having completed the operation.

Flight Lieutenant Charles (Wyn) Cartwright and his men in 6225 BD Flight came across a large number of V1 warheads at a bomb dump in Schleswig Holstein. These all had to be disposed of by demolition. By the end of the war some 4,370 tons of V-weapons had been dealt with in this way. (*Author collection*)

Later that same day Flight Lieutenant Cox of 5139 BD Squadron reported yet another V1 to be visited and treated with the 'J' process. He asked Flight Sergeant Martin to meet him after dealing with the Doel bomb and requested he bring his team with equipment that included timber and rope. The circumstances of this bomb are not known but Martin and his party returned on 4 March having successfully dealt with it.[54]

As March went on 6210 BD Flight were still dealing with V1s too. On the 13th they were at Bercham on the southern outskirts of Antwerp where the Germans had a V1 dump. Two of the warheads were in such a dangerous condition they could not be moved and were destroyed in situ. Another two warheads were removed to the Flight's bomb dump.[55]

By the end of that month the Allied BD teams had dealt with quite a number of V1 warheads. It is recorded that Flight Sergeant Beer, working in Flight Lieutenant Cartwright's 6225 BD Flight, undertook demolition of two on 26 March in the Eckernforde area. A further three V1s were dealt with on 28 March and another on the 29th.[56] One V1 that had to be found had crashed unexploded near Bromskirchen (Neuludwigsdorf). The wreckage of this one was taken to Neuludwigsdorf and was said to have been stored at the forester's house for a while. Often wreckage was picked up by local farmers, and put to good use.

Also in this locality was a railway used for transporting V-weapons. On 29 March 1945, a train carrying V2s came to a halt near Bromskirchen. It was a very long train, requiring both an oil-fired locomotive and a coal-powered one to move it. The coal-fired one needed to be replenished with water, hence the reason for stopping. The train crew felt very vulnerable, so they moved on half the train using the oil-fired locomotive, while the rest was left to be replenished. The crew were right to worry – soon, a tank of the American 3rd Armoured Division arrived at the scene. The tank team fired on the train and put the locomotive out of action, the crew fleeing. On inspection of the captured train the Americans found 9 damaged V2s as well as a number of warheads and a scattering of parts from another V2. The rest of the train was later found in a tunnel in the Brilon Forest and contained a further 12 warheads and various V2 component parts.[57]

A factory nearby at Hatzfeld was also captured and found to contain a number of intact V2s.[58] The media at the time took newsreel footage of some of these trophies and there was a lot of interest generated. Eisenhower even

The demolition of leftover ordnance, including V-weapons, involved some pretty big explosions. Bill Morton-Hall commanded 6205 BD Flight while it was in Germany. He recalled that they produced a mushroom cloud almost every other day and described how they used a small concrete shelter 2,000 yards from the demolition site. The photo shows the result of detonating 100 tons of ordnance at once. According to Bill, 'At the moment of detonation it was necessary to bend the knees, keep mouth wide open, hands over ears and whatever you did, you did not rest against the shelter wall.' (*Author collection*)

Europe 113

American forces captured this train loaded with V2s, pictured at the entrance to a tunnel in the Brilon Forest. (*The National Archives: Ref. AVIA 6/25653*)

This is the rear end of the V2 warhead. The hole in the centre was for the base fuze. The hole in the eleven o'clock position connected the alcohol tank with the hole in the nose piece (see next photo). (*The National Archives: Ref. AVIA 6/25653*)

The hole seen in the warhead is part of the V2's fuel delivery system. The alcohol in the fuel tank is required to be under pressure. A tube runs from this hole through the warhead to the alcohol tank. Because of the speed the rocket travels at, the air passing through this hole pressurizes the tank. At high altitude however, atmospheric pressure is lower, so this tube has a valve that automatically shuts. Nitrogen is then fed into the alcohol tank under pressure. (*The National Archives: Ref. AVIA 6/25653*)

visited the site. The rockets were subsequently taken away to Antwerp for onward shipment to the USA for further investigation and tests.

The destruction of abandoned V-weapons and their warheads continued apace. On 8 April 1945, 6210 BD Flight removed two V1 warheads from Knokke to Brussels for demolition by 6206 BD Flight. On 11 April, Sergeant Retallick and his party from 6205 Flight demolished a quantity of munitions at Evere on the eastern outskirts of Brussels that included two V1 warheads, more than likely the ones 6210 Flight dealt with.[59]

Four V1s were sent back to the UK for investigation at this time – Flight Lieutenant Booth and Corporal Brown began to render the missiles safe on 27 April. The following day Warrant Officer Dyer and Corporal Brown completed and certified the operation.[60]

On 23 April at Lomers, near Rohr, 6213 BD Flight (5130 BD Sqdn), based at Nieder Mendig, destroyed two V1 warheads. The explosive in them was rapidly deteriorating so they were destroyed where they were found by electrical demolition.[61] They destroyed two more on 25 April, at a rocket site situated between Lommersdorf and Rohr.[62] The problem of explosive deteriorating was beginning to be critical.

The Germans had left some of their sites in a real mess – one example is Eggebek. 5132 BD Squadron had only arrived on the Continent on 16 May. On the 23rd the OC visited Eggebek to be greeted by 6208 BD Flight, who had arrived the day before.[63] They found 21 V1s in nearby woods – all had had their motors demolished but the warheads left intact. Most were considered to be in a dangerous condition, as the fuzes had been removed and the picric pellets were open to the weather.[64]

Still in May, 6225 BD Flight based at Kropp undertook some major demolitions. On the 13th they destroyed 12 tons of explosives, including 6 V1 warheads. Local farmers were a bit upset, claiming they had not been warned. A small peat fire was started but as a ditch surrounded the field it was contained and did not present a problem.[65]

On 19 May, 6201 BD Flight cleared 3.5 tons of miscellaneous German munitions, consisting of mortar bombs, grenades and panzerfausts, from what was described as a V1 and V2 site at Dearbusch. A search of the area revealed that all the crashed V1s from here had exploded. The flight was at another V1 site a few days later on the 23rd, at Buchel (25 miles south-west of Koblenz). Here they dealt with a V1 warhead and 4 damaged 50kg bombs and other munitions. A total of 1.5 tons was demolished.[66]

On 27 May demolition continued at Kropp and Flying Officers Pink and Allison visited an ammunition dump to check out some V1s.[67] These dumps were becoming major hazards and their destruction was a time-consuming job for the Allies. Flight Lieutenant J. M. O'Connor of 6201 BD Flight records that on 5 June 1945 approximately 19 tons of enemy and Allied explosives was demolished. This included 6 crashed and broken V1s (5 fuzed).[68] Two days later 2 more crashed V1s were demolished near Tondorf. In the same area 6201 BD Flight destroyed 3 V1 warheads on 29 May at a V1/V2 site. They also dealt with a V1 warhead that was lying by the side of the main road from Trier to Cologne (Route 51). It was towed by a lorry some 400 yards away

Lieutenant James Patchell, CO of the 75th BD Squad, next to an unexploded V1 that had embedded itself vertically in the ground. (*T. Dennis Reece*)

A V2 near Celle deliberately destroyed by the Germans to avoid it being captured by the Allies in April 1945. (*Author collection*)

Photographs printed in *Evidence in Camera – Flying Bomb Issue* of 24 July 1944, with the caption 'They don't all reach England'. (*Author collection*)

A V1 on display marked 'safe for transit' by 6228 Bomb Disposal Flight. (*Author collection*)

from the road and demolished in a field.⁶⁹ In June 1945 another 8 unexploded flying bombs were dealt with by 6201 Flight.⁷⁰

At the same time, the Allies were hunting for complete examples of V-weapons that they could recover and ship back to the UK and USA in order to exploit the technology. It was the Air Technical Intelligence that needed V-weapons recovered, but of course the job of finding them and making them safe was for the bomb disposal officers.

In May 1945, 6228 BD Flight (5138 BD Squadron) based at Twente airfield, examined a large number of V1 and V2 missiles with a view to recovering them. In the official report at the time Squadron Leader A. S. Dykes of 5138 BD Squadron regrets that some warheads which others had considered safe could not be certified for transit back to the UK. He cites a previous experience (for which they were indebted to their liaison with the Royal Engineers), and therefore the quantity recovered was less than anticipated.⁷¹ He noted that the Royal Engineers had suffered six casualties through certifying similar warheads safe. The tragic event referred to happened to a Royal Engineers BD squad on 30 April 1945 close to the village of Holten. No. 24 Bomb Disposal Company, under the command of Major Tom Sharman GM, was tasked with dealing with a V2 there. In the Dutch-language book *Holten at War* there is an account from 1981 by Jim Brockbank, a British ex-serviceman, who was a witness to what happened that cold and rainy day. Brockbank had been Staff Sergeant with 147 Coy RASC, a bridge laying and repair unit. At 1100 hrs he was heading towards the village of Holten in a Dodge truck driven by his comrade, Fred Court, on their way to Arnhem. As they got to the T-junction of Rijssenseweg and Markeloseweg they saw from the left a bomb disposal lorry which to their surprise was towing a V2 warhead by a length of chain. Brockbank had seen the warhead before – it had earlier been on the back of a broken-down German lorry. Now it was being dragged along the road with some soldiers around it.

Brockbank and Court's lorry turned towards the village, but as they drove on there was a big explosion. Court was wounded in the back and Brockbank in the face. Locals took them into a house before they were taken to hospital at Deventer. Jim Brockbank was operated on but lost his sight in one eye.

Six people were killed in the explosion. Five were members of No. 24 Bomb Disposal Company: Corporal J. P. Coyne, Sappers G. E. Wareing, J. S. McWhinnie, L. Cotton and Driver J. E. Harris. They were buried at Almelo cemetery, except Coyne, for whom there were no identifiable remains. It was said by a witness that he had been sitting on the warhead at the time it detonated. His name is instead on the Groesbeek Memorial to those with no known grave. The sixth person killed was a 23-year-old evacuee girl, Corrie Lepelaar. Another woman, who was pregnant, was injured by the blast.

Major Tom Sharman (*Lionel Meynell*)

Lieutenant Brian Richards (*Lt Col Eric Wakeling Ret'd*)

Fifty years to the day later, on 30 April 1995, former OC 24 BD Coy Major (Retd) Tom Sharman GM, 2IC Major Bryan 'Ricky' Richards GM and a Mr Cotton (brother of one of the Sappers killed), attended an unveiling of a memorial near the spot. Tom Sharman's hostess at the event turned out to be the daughter of the injured woman, who had given birth ten days after the explosion.[72]

Incidentally, both Tom Sharman and Bryan Richards were awarded the George Medal, Sharman for working on three different delayed-action bombs at factories around Birmingham, one of which was also fitted with a booby-trap fuze sensitive to the slightest vibration.[73] Lieutenant Brian Richards worked on a bomb for 18 hours in a flooded shaft at the Yorkshire Grey Dance Hall, in London's Eltham Green. This bomb's fuze 'ticked' intermittently.[74] For this action General Taylor recommended him for the George Cross, his efforts described as 'an outstanding example of cold blooded courage and determination'. He was subsequently awarded the George Medal.[75]

No. 1 Dutch Bomb Disposal and Mine Clearance Company at Kruisland, 7 April 1945. Formed in early November 1944, this BD company rendered safe and disposed of quite a few V1s. Between 8 March and 16 June 1945 Section No. 3 alone is reported as having dealt with the following unexploded flying bombs:

8 March 1945 – Ossendrecht – this V1 crashed on the night of 7/8 March at Calfven. One no. 80 fuze was still found to be working.

14 March 1945 – Huybergen – a V1 crashed 11 March at 1800 hrs at Demerstraat. It was defused and the explosives destroyed in several detonations.

29 March 1945 – Ossendrecht – this one crashed on the morning of 29 March on farmland at the Nieuwhinkeleroord Polder. It was defused and the explosives detonated in small amounts.

6 April 1945 – Steenbergen – this V1 also crashed on the morning of 29 March at the same polder. It, too, was defused and the explosives detonated.

12 April 1945 – Kruisland – this one crashed on the afternoon of 11 April on farmland 100m from the Nieuwe Biezenweg. It was defused and the explosives detonated in small amounts at a safe location.

23 May 1945 – Kruisland – a V1 crashed on farmland near the Ringweg. It was defused and the explosives detonated.

15 or 16 April 1945 – Steenbergen – this V1 crashed in the Julianapark near some bunkers and broke into three parts. It was defused and the explosives detonated in small amounts at a safe location.

(*Stichting Geschiedkundige Verzameling EOD, Netherlands*)

Another incident in which people were killed by the unexpected detonation of a V2 warhead occurred on 4 December 1944. A V2 rocket launched from Hellendoorn, Holland, crashed back to earth without exploding about 5km away in fields near the town of Luttenberg. Many saw the rocket come down and rushed to where it had impacted. An eye-witness report (see website www.V2rocket.com) recalls a big hole in the ground which still contained a fierce flame burning. Around the edge of the hole many curious locals were gathered. Suddenly the remains of the V2 exploded. The eye-witness, Mr A. Kleine-Toereers, was close to

No. 1 Dutch Bomb Disposal and Mine Clearance Company with another defused V1. Flying bombs would continue to be found in the Netherlands for many years after the war. In the 1970s and 80s it is estimated that the Dutch Army dealt with 20 incidents involving unexploded V1s (see Chapter 6). (*Stichting Geschiedkundige Verzameling EOD, Netherlands*)

A Dutch Captain (possibly Captain Van der Sleesen) with another Allied officer. Notice the V1 they are leaning on still has the armed fuzes in place. (*Stichting Geschiedkundige Verzameling EOD, Netherlands*)

German prisoners of war working on a crashed V1 near Oss in the Netherlands. It appears they are attempting to peel back the warhead's casing in order to remove the explosives. (*Nederlands Instituut voor Militaire Historie, The Hague*)

An unexploded V1 in the Netherlands. The man on the left is Lieutenant Aalpol. (*Stichting Geschiedkundige Verzameling EOD, Netherlands*)

the explosion and was pushed to the ground. It was ten minutes or so before he could recover enough to leave the site. As he did so he met two acquaintances who pointed out that his clothes were smouldering and as for his trousers, only the top of them remained.

A nurse who had been at a farm close to where the V2 came down recalled that the rocket exploded approximately 15 minutes after impact. She grabbed a first-aid kit and headed for the scene. There she tended to the injured and was soon joined by German ambulances. In all 19 people died as a result of this incident, the youngest being a couple of children only ten years old. A memorial to the victims has been erected on the spot.

Any parts of V-weapons that were discovered were of interest to the Allies. Sergeant Cecil Brinton had served with No. 2 Bomb Disposal Company in London. Later, while he was serving with 211 Field Park Company, the Canadians requested his assistance. They had come across a train in an area of Holland they were controlling, consisting of eight trucks carrying V2 parts. These would obviously be ideal for research purposes but there was a problem – the whole train was booby-trapped.

Sergeant Brinton found each truck had a number of explosive charges rigged up with trip wires and instantaneous fuze wire. Different ignitors had been used – some would detonate the charge if the wires were stretched, while others would detonate if they lost tension (ie were cut). The first and last truck also each had a 500kg bomb attached, for good measure. Brinton was able to render safe all the charges including the two big bombs. For his efforts, on 21 June 1945, the *London Gazette* published his award of the British Empire Medal.[76]

Some complete examples were successfully collected. At the request of Wing Commander Hood of the RAF Mission to Belgium, Flight Lieutenants Cook and Yorke-Smith of 6229 BD Flight visited an exhibition on 16 August 1945 being run by Hood in Brussels, in order to examine a complete V1. After scrutiny it was found to be in a safe condition for transit.[77]

However, transit did not always take place as planned. Another V1 was inspected by Sergeant Slater and his party from 6229 BD Flight at Lokeren and loaded on a trailer on 6 September in readiness

Sergeant Cecil Brinton. (*Lt Col Eric Wakeling Ret'd*)

Europe 125

Photo taken June 1945 at Oyle, Germany (between Bremen and Hanover). An RAF bomb disposal squad (6228) have marked this V2 warhead as safe to transit. (*Author collection*)

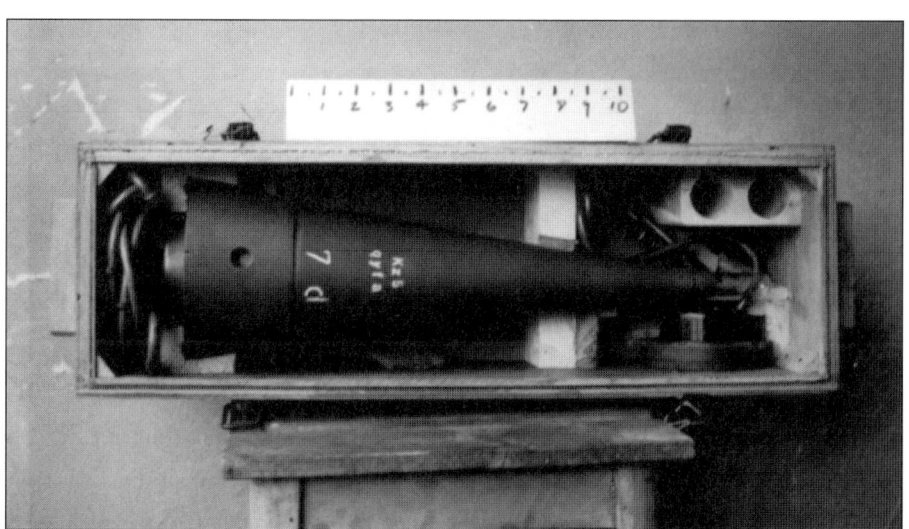

Nose piece from a V2 found still in its packing case. At the tip was a glass cover that went over a simple crush-type switch. In addition, the nose fuze incorporated two trembler switches at right angles to one another. Another fuze located at the base of the warhead was found to have a similar set up. (*The National Archives: ref AVIA 6/25653*)

for being shipped back to the UK.[78] Again a trip was made to Lockeren by 6229 BD Flight personnel on 20 September, when Flight Lieutenant Cook and Sergeant Westbrook were tasked with removing the warhead.[79] Corporal Phillips and party collected it the next day. On the 24th, Corporal Hewitt and Leading Aircraftman Murphy took the vehicle loaded with the V1's warhead and other miscellaneous bombs (eighteen 500kg bombs, one 2000lb HE bomb and two French 40kg bombs) to Bourg Leopold for demolition, which took place the next day. Flight Lieutenant Cook had gone back to Antwerp on the 23rd to check on the V1 awaiting shipment to the UK, but was advised by Squadron Leader D. Dykes that he should hang fire, as another specimen was being arranged. Later, he was told that the original V1 was now to be destroyed, as three others were available. Cook was asked to certify these three safe for transit to the UK, but when he went back to Antwerp on 7 November to do so he found there were actually five V1s there (four from the Royal Engineers). He could see that the warheads appeared to be empty and the warrant officer was instructed that further information as to where the bombs

A V1 at a 9th Air Force airfield in France. Note the shaped piece of wood that could be part of the storage packing, indicating that this bomb might not be one that crashed. (*Author collection*)

had come from and who had certified them safe for transport was necessary before authorization could be given for transit to the UK. A case of too many cooks![80]

One of the largest V-weapon facilities found was the underground 'Mittelwerk' factory at Nordhausen. Over 4,500 V2s were produced here, using slave labour in atrocious conditions – it is said that more people died

The same bomb being readied for transporting to the USA. (*US Air Force*)

making the V2s than were killed by their explosions. The Americans had first discovered the factory on 11 April 1945. During May they removed tons of V2 parts and sent them back to the USA. At the end of May the British arrived to pick over what was left.

On 6 June, 6203 BD Flight was instructed to proceed to Nordhausen and make contact with Colonel Warner (USA). They were tasked with removing as many V2 rockets as possible. Next day a squadron technical officer with 6212 Bomb Disposal Flight went directly to Nordhausen. The same day 6203 BD Flight arrived from Detmold. The strength was 4 officers and 38 other ranks. Later this was bolstered to 60 with the arrival of 2 Cole Cranes and 8 Queen Mary trailers.[81]

On 8 June things were starting to happen and everyone present was brought together. The site consisted of two main tunnels connected by 40 galleries quarried into a limestone hill. Artificial light and air-conditioning was operating within the tunnel complex. Army units of the REME and RASC had arrived two days previously and were now loading steam units and lifeline pumps into railway trucks. A German electrician had also been co-opted and an almost complete layout of a V2 was now present. One working party loaded V2 fuel tanks on to Queen Mary trailers and another party loaded the numerous small components into three-ton lorries which were able to enter the tunnel complex from the opposite end to the main working party. On completion of loading, the vehicles were driven to Kassel, 240km from Nordhausen, where they were to be put on the railway to Cuxhaven.

Later that day more vehicles turned up to help with the task. Transport strength was now 10 three-ton lorries, 8 Queen Marys, 2 Cole cranes and 6 ten-ton lorries. Loading continued on 9 June, it being noted that there were 18 types of alloy tubing of the same diameter for the V2's venturi unit. Acting on American intelligence, a private house was raided in Bleicherode with the result that three boxes of valuable control equipment were recovered. On the 10th, the loading parties were still hard at it loading fuel tanks and fuselage sections. Instead of sending the laden vehicles to Kassel it was now proposed to send them to an airfield at Gottengen (inside British territory), as it was closer. On the morning of 11 June work reached its peak: 8 Queen Marys, 6 ten-tonners and 4 Bedfords were loaded. As an experiment 6 half fuselages plus one fuel tank were loaded on to a Queen Mary, which arrived at Gottengen without mishap. The Bedfords continued to be loaded with the alloy pipeline assemblies. Work continued the next day, the men feeling the pressure to get as much done as possible. To date they had loaded 64 fuel tanks and 53 half fuselages. Loading up railway trucks with five V2s at Klein Bodungen was finished, now they awaited a steam engine to take them away.

As yet no gyro units had come to light at Nordhausen. Five V2s with three launching units that were found in the grounds of the repair shed of the salt mine were moved by rail to the siding at Niedersachswerfen. The mouth tunnel party spent the day in loading air bottles, of which there appeared to be fewer than 100. A visit was also paid to Gottengen airfield where an empty hangar, complete with manually operated overhead crane, had been set aside as storage space. The main party continued loading fuel tanks and fuselages.

On 14 June, very heavy rain made the outdoor work of the Coles Cranes impossible, as they were unable to negotiate the boggy approach to the stack of fuselages. An alternative method was therefore employed in which 12 ex-storm troopers from the prison camp were forced to do the job – with what is described as 'an hilarious audience of displaced persons' looking on. By now the total material recovered stood at 47 alcohol tanks, 74 oxygen tanks and 143 half fuselages. It was thought that the supply of alcohol tanks might have been a problem as there were not many left. A dozen more three-ton tenders arrived at Nordhausen on 15 June, to supplement the transport. The mouth tunnel party brought the small components task to an end after loading items such as nuts, bolts and washers, and by the 16th the operation was virtually complete, except that the supply of alcohol tanks appeared to be exhausted. Another 6 were found at Klein Bodungen, which raised the number to 105 – the total recovered so far was 205. That evening it was decided that the drivers of the Queen Marys could be stood down the following day for a well-earned rest.

On 18 June, while searching through an unlit gallery, 25 more alcohol tanks were found and it was decided to attempt to get them out. Four three-ton lorries were driven into the tunnel and one tank was loaded on each. On attempting to drive out at the other end it was discovered that the diesel engine that had been used for shifting the railway trucks had been left midway through the tunnel and could not be passed. There was no engine driver handy but someone succeeded in starting the motor. It was unfortunate that a member of the party then leaned on the brake handle, and the engine set off to a flying start! However, the brakes were hastily applied and no harm was done. A more cautious investigation resulted in a squadron leader and the BD squadron technical officer successfully driving the engine out of the tunnel and subsequently all 25 alcohol tanks were recovered.[82]

Despite both the Americans and British removing V-weapon parts in huge quantities from the Mittelwerk site, to this day remnants of the weapons can still be seen. Within the tunnels a museum now displays some of these artefacts.

This photo is believed to show Royal Engineers of No. 23 BD Company in the process of recovering a V2 from Germany in May 1945. The missile was brought back to the UK for training purposes and is possibly the one passed to the Royal Engineers Museum at Chatham for public display in 2012. (*Author collection*)

Much of the information on recovered V-weapons in this chapter has concentrated on the efforts of the RAF, but of course the Army and Navy were also involved. The Royal Navy from the start was interested in obtaining information that would help it deal with any V-weapons it might come across. Back in January 1944 the director of the Naval Unexploded Bomb Department issued a directive to bomb safety officers in 17 ports around the UK, telling them that no attempt at examining any new type of missile should be made if it entailed movement of the weapon or its components.[83] The BSOs were to inform the director who would then arrange for an examination using personnel who were best informed on the subject. There is no record that the Royal Navy ever had to deal with any unexploded V-weapons around British shores, but Europe was another matter.

The Navy was tasked with clearing the European ports of mines and booby-traps that had been left by the fleeing Germans in order to disrupt the supply operation to the advancing Allies. Every inch of the bottom of the docks had to be checked to ensure they were free from hidden charges. Bear in mind Antwerp alone, pre-war, was the world's largest port. An area of nearly 8.5 million sq ft had to be searched and the job was made more difficult by the fact that during the war years the docks had not been properly dredged, if at all. The divers were nearly always working by touch alone and in some instances were literally up to their necks in mud at the bottom of the docks. It

Sub Lieutenant Arthur Russell MBE. (*Martin Russell*)

Able Seaman Raymond 'Brum' Maries BEM, at the Royal Navy Divers Reunion in 2008. He was Mentioned in Dispatches for his part in clearing the European ports of UXBs. (*Minewarfare & Clearance Diving Officers' Association*)

was even recorded that in one basin the silt was deeper than the divers, which meant it covered their heads and choked the air outlet valve on their diving helmets, preventing them from breathing. Sometimes the water was so cold a diver could only stay down for ten minutes before his hands were too numb to feel anything properly – a dangerous condition when working on bombs, and relying on touch alone.

It was in these circumstances that on 14 March 1945, Naval Party 1572 was called to find an unexploded V1 that had come down in the Royal Sluice at Antwerp Docks. An Irishman, Lieutenant Commander H. J. Horan, with Sub Lieutenants Arthur D. Russell and R. Blyth working under him, led the 'P-Party'.[84]

One of the divers in the search party was Able Seaman Raymond 'Brum' Maries. Looking back in 2010, he recalled that Antwerp was on the receiving end of many V1s while he was there. The P-Party had Royal Marine drivers who, on one stretch of open road to the docks known as Doodlebug Alley, would really put their foot down. He was of the opinion that they had been tasked with looking for the first unexploded example and knew that once recovered it would be sent back to the UK. Unknown to him others had already been recovered, of course, though complete examples were always required for intelligence gathering purposes.[85]

The Antwerp Docks V1 was found by a man on an adjacent line to Maries, several feet below the surface and deeply embedded in the mud. Able Seaman Robert Gribben, a peacetime regular, volunteered to partner an unnamed officer for the operation and, digging with their bare hands, they managed to burrow in from each side of the fuselage in order to pass a strop around the bomb. It was exhausting work which took them nearly a week.[86] Much of the time they had to work in an inverted position.

The doodlebug was eventually lifted out virtually intact. Ray Maries remembered that in the tail end they found propaganda leaflets. He recalled the scornful jokes they made about the leaflets, one of which advised 'How to repair bomb damage', while another explained 'How to cook spam'.[87]

A report of the time describes Gribben (who was awarded a George Medal for this job), as 'a diver who knows no fear and has carried out his searches in a very thorough manner. His diving hours rank with the highest in his party and he was always amongst the first to volunteer for any exceptional work. It is impossible to speak too highly of his work in this and all other operations. He has constantly shown a sense of pride and responsibility in all his work with initiative and exceptional stamina both physical and mental. His courage and devotion to duty is of the highest order.' Gribben continued to be involved with Navy bomb disposal, earning himself a Mentioned in Dispatches in the 1952 New Year's Honours. Russell was awarded an MBE for his work in clearing the ports, which entailed him continually diving under extremely hazardous and dangerous circumstances during which it is noted that he displayed the utmost coolness – the threat of sudden death was always present.[88]

Lieutenant Commander Peter Keeble was a salvage diver who had experience of bomb disposal that included disarming mines in the Mediterranean. On the very last page of his book *Ordeal by Water*, he makes the briefest mention of having to deal with a submerged V2. This event apparently occurred right at the end of the war when Keeble was involved in 'a dreary tour of the war-shattered ports of Europe'. He recounts that the 'tour' was 'slightly enlivened by a brush with an unexploded V2 on the bottom of the River Scheldt', but unfortunately does not go into any further detail.

There were other V-weapons that came down unexploded in rivers and the sea, a couple of which have been mentioned already. A number of aircraft carrying V1s were shot down and crashed on water, and as the V1s attached to them would not have been armed, they would not necessarily detonate. An example is the Heinkel III A1+BH that was shot down with its V1 still attached on the night of 24/25 November 1944. It crashed into the North Sea a few miles off the Dutch coast to the west of the town of Alkmaar. The body of the pilot, Feldwebel Kurt Hillmann, was washed ashore a couple of days later.[89]

Another aircraft that crashed while carrying a V1 was a Heinkel 111 of 9/KG 53 that came down on 5 January 1945. It had taken off from its base at Eggebek, in Schleswig-Holstein near the German/Danish border, but before it left German airspace it ran into problems. The port engine failed and the aircraft went into a gentle dive. There was no time to jettison the V1 or the fuel before the aircraft crashed in a wood at Westerlangstedt (south of Eggebek). The V1 was thrown clear of the burning wreckage on impact and, luckily for some of the crew, did not explode. Though the gunner and radio operator were killed, the pilot survived with slight head injuries and burns, the observer was thrown out and fractured his skull, and the flight engineer sustained broken ribs. These three surviving crew were taken to hospital at Schleswig.[90] Who knows how many other V1-carrying aircraft crashed into the Channel or North Sea under similar circumstances?

Chapter 5

Tools of the Trade

By the time V-weapons were encountered in 1944, the Allied bomb disposal organizations had refined their process both in the detection and reporting of UXBs, as well as in their disposal. Technology had come a long way since the beginning of the war and both sides had always to consider new threats and ways to deal with them. Having the right tools to hand, ones that were proven to work, made the life of those engaged in bomb disposal much easier. However, the right tools did not always exist, or if they did they were not always readily available. Sometimes a jeep toolkit or even a bicycle repair kit had to be used to disarm bombs when nothing else was to hand. Certainly in the early days tools were virtually non-existent.

When the war began aerial attack was already recognized as a great threat and Britain was preparing its citizens with the construction of air raid shelters and other defences. Little thought had been given to unexploded munitions. It was almost assumed that they could be literally dragged away and disposed of with relative ease. This was far from the case. Bombs dropped from the air could penetrate the ground to quite some depth and the angle they struck and obstructions they hit could deflect their path through the ground. Much digging was often required just to expose a bomb, before the technical aspects of disposal could even be considered. This was nearly always the case, and V-weapons were no different.

Locating the precise position of a UXB or V-weapon warhead was one of the first priorities. Bearing in mind they might contain some sort of time-delay fuze, finding them had to be done very quickly and efficiently yet with great caution. Probes were developed that could help detect their path through the ground. Of course, probing for a bomb could be a dangerous undertaking. If a fuze was sensitive to vibration then too aggressive a jab might cause it to detonate. An officer's swagger stick was just not up to the job, not being nearly long enough. A proper bomb probe could be extended by adding 4ft-long sections, as with a chimney sweep's brush, and had a pointed end that could penetrate the earth easily. On the other end there was a handle much like the one on a garden fork. Lieutenant Colonel Eric Wakeling commented that they rarely found the actual bomb with one of these probes, but often

Probably the most important tools utilized by the men of bomb disposal were the already successfully disarmed weapons. These training aids enabled men to study and handle the actual equipment and gave them a confidence in their own abilities in a way that the text books could not. Blindfolds could be worn while practising the disarming process in order to sharpen the senses. This V1 is shown among a few other examples of captured munitions at the US Navy's School of Bomb Disposal, at Washington DC in 1945. (*Author collection*)

found the tail fins that had been ripped off as the bomb passed through the ground.[1] They were good though for finding the angle the bomb struck, from which one could estimate where it had actually come to rest.

Once the rough location of a bomb was worked out, the next job was to dig it up and identify properly the fuze, or fuzes, fitted. Again, caution was a priority whether it be a V-weapon warhead, a bomb or a mine.

Digging was generally done by hand and usually by men of the lower ranks. However, in some cases the officers would get involved. In the early days of mine disposal the officers were issued with a child's seaside wooden spade, as that was the only non-magnetic digging tool that was immediately available and safe to use around magnetic mines. Later, proper spades and shovels with blades made from non-magnetic phosphor-bronze were issued to bomb disposal units.

Experiments were conducted in 1941 by the Road Research Laboratory to see how great the risks actually were for the men doing the digging. Remember that the men could often be digging towards a bomb that was fitted with both a time-delay fuze and a fuze sensitive to vibration. In the experiment it was found that a type 17 clockwork time-delay fuze in a sensitive condition (stopped but ready to restart), and a type 50 anti-handling fuze, could be activated if a single blow with an ordinary spade was made less than two feet away from the buried bomb.[2]

Digging was thus a major part of bomb disposal work and those doing it did not always get the credit they deserved. Some bombs were found as deep as 65ft.[3] The V2 warhead at Paglesham was recovered from a depth of 37ft. The excavations had to be quite wide too, in order to build platforms to move earth to the surface in stages, and because a bomb could change direction under ground. Sometimes a number of shafts were needed to find one bomb.

As the depth of an excavation increased, so the sides would require shoring up. It was not just the bomb or warhead that was a hazard. Early in the war a number of men died as a result of excavations collapsing and burying them alive. One such incident occurred at Lowestoft, Suffolk, on 15 April 1941, when the hole collapsed, completely burying Lieutenant Hoare of 22 BD Company. He was dug out before he suffocated and taken to hospital. The bomb was recovered and Lieutenant Hoare discharged himself from hospital and set about defusing it. As he did so it exploded, killing him and another man, Corporal Gibbs.[4]

Again in the early days, shoring up shafts with wood, assuming it was available, was a little haphazard. By 1942 things were more organized, with standardized methods of working and materials. Planking and piles were introduced as part of a BD squad's kit, greatly reducing the risks. The Royal Engineers specifically had a lot of carpenters and other skilled trades in their ranks and they soon found their way into bomb disposal to help with the engineering side of things. In fact when units had a quiet spell with no bombs to work on, they could often be found employed on construction projects, such as building beach defences or putting together Nissen huts.

Despite better training and equipment, accidents did still occur. One officer, Captain Henry Grover GM, was killed whilst attending a demonstration of shaft digging and timbering in July 1942. Grover was a big man, an ex-First World War Royal Flying Corp pilot, and while he was inspecting the top of the shaft, the timber gave way and he fell to the bottom of the 15ft hole. He was killed by earth and timber falling on top of him.[5]

Bombs in timbered shafts were not generally blown up in situ. Timber was considered part of the unit's equipment and was to be reused job after

Compare the cramped conditions of the unsupported shaft in the photograph on the left with the timbered shaft on the right. In the first photo a short tunnel, or 'sap', had been excavated, as the bomb was found away from the centre of the shaft. Having earth overhanging a bomb that was being worked on was not ideal – a clod of earth falling at the wrong moment could prove fatal. (*Author collection*)

job. As it was imported it was in relatively short supply and much was lost on account of convoys being attacked and bombing of the docks – as a result, it was never left lying about by the bomb disposal squads for fear it would be stolen.[6] Occasionally, steel piling was used where there was a lot of water or very unstable ground but this required heavy plant equipment in order to lift the piles and force them into the ground.

Flooding of excavations could be a real problem, causing the walls of the shaft to collapse or the bomb to sink out of reach. Good pumping equipment was needed in such circumstances. It might also be the case that the ordnance had fallen in a river or tidal region, where good fitting piling and pumps were an absolute necessity.

The British forces were issued with a number of different types of pump. Sometimes they would also make use of other people's, such as the Fire Brigade's. The pumps used generally were the Skyes diaphragm type that could cope with muddy water, and the Winget centrifugal type, which was better with clear water. Another way to lower the water level in the bottom of a shaft was to sink tubes into the surrounding ground at intervals. These could then have pumps attached to remove the water. Often pumps were simply not up to the job and bombs were abandoned as they sank deeper in the soggy ground.[7]

In 1996, in response to a Parliamentary Question, the Armed Forces Minister Nicholas Soames released a list detailing the location of over 40 known German unexploded bombs that had been abandoned in the London area.[8] From the addresses, it is obvious that wet conditions would have been the reason that many of them were never recovered. Below are a few of those listed:

Dagenham – sewage farm near River Rom, eastern end of Western Avenue
Dagenham – river bank by sluice gates, Horseshoe Corner, 50 yards east of sluice gates
East Ham – Sprengbrand sludge troughs sewage farm
East Ham – rear of Creon and Silley weirs, 20ft to north side of creek in allotments

In 1996 nine different London cemeteries were known to have an abandoned unexploded bomb still buried. This Stepney graveyard had one that was situated behind 33 Ropery Street (house just visible beyond the wall). (*Author collection*)

Enfield – East of River Lea, 30 yards from Lea Bridge
Leyton – Leyton Marshes, rear of Latham's timber yard
Leyton – 100 yards south of Lea Bridge waterworks, on marshes
Poplar – LMS goods yard, West India Dock
Stoke Newington – number 9 filter bed, Metropolitan waterworks, Green Lane, N16
Tottenham – Marsh Lane
Waltham Cross – in old River Lee, near King George Reservoir
West Ham – in River Lea, near 339 Carn Rd, Stratford
West Ham – Lloyds Rubbish Shoot, Marsh Gate Lane
West Ham – number 3 pit, Shellmar Wharf, Silvertown, E16
West Ham – Harland and Wolff, Manor Way
Crayford – allotment in marsh, Bourne Road, near tidal stream

A big muddy hole was dug at Holywell Park, Ipswich, in January 1941 to recover a huge UXB. The 1,800kg 'Satan' bomb measured 8ft long and more than 2ft in diameter. Heavy plant was used to dig a ramp down to the bomb as the ground was so wet. Pumps were in place to try to keep the hole dry. It took a few days in miserable conditions, but the bomb was finally extricated and put on public display. (*Photos: Author collection*)

Crayford – 150 yards north-west of the junction of the rivers Cray and Darenth
Erith – riverbank by Thames, Crossness, Abbey Wood
Erith – disused sandpit off Fraser Road, River Walk
Erith – can site, Harrow Manor Way, Belvedere Marshes
Erith – 4 at old LCC sewage farm, Crossness plot 119/120, East Road, Belvedere
Feltham – 3 in river bed, Longford River
Heston – bed of canal, Northumberland Wharf, Brentford
Richmond – mid-Surrey golf course in ditch between fairway and Thames
Sunbury – footpath, Pharoah's Island, Shepperton
Wimbledon – Wimbledon sewage works

This list caused quite some concern, especially as some of the bombs were in the gardens of residential homes. Technology continued to advance of course, so a number on the list were in fact later recovered. Those remaining are considered not to be of any great risk.

Pulling a bomb from an excavation was another risky operation and required tools and equipment suitable for the job. One essential was a set of 'sheerlegs' or a 'gyn'. A gyn had three legs and could only move a bomb vertically. Using these a bomb at the bottom of a shaft could be rotated to make it easier to maneouvre. The sheerlegs had two legs and enabled a bomb to be moved sideways, for example into the back of a lorry, rather than having to roll it up a timber ramp. A V1 or V2 warhead weighed a ton, so they were not easily manhandled into a truck.

An accident involving sheerlegs occurred in July 1943, while the Royal Navy were recovering an unexploded mine that had washed up on the shore in a small cove at Frenchmans Bay, near South Shields. It was disarmed without incident by Lieutenant Blackmore RNVR. However, the next day it was hauled to the top of the cliffs using a block and tackle suspended from a pair of sheerlegs. Unfortunately the ground sloped away from the cliffs and as the mine was lowered back down, it tipped and pulled the sheerlegs over centre. A warning was shouted but Leading Seaman James Robertson appeared to slip and was struck on the head by the falling sheerlegs. Lieutenant Dowden, who was next to Robertson, was also hit but only suffered minor injuries. A medical officer from a nearby army battery came to help but Robertson was already dead.[9]

Squadron Leader Dinwoodie, an experienced bomb disposal officer, wrote a report on the effectiveness of equipment employed. In it he stated that the sheerlegs was a very useful tool as it could be used to suspend a bomb

A sheerlegs or a gyn, as seen in the photo above being used by 218 BD section, would be used to hoist a bomb from its hole. It could then be swung on to the back of a lorry, so long as a lorry could gain access to the top of the shaft. Only some of the lorries in use by British BD units had the 'luxury' of a crane on board – when the companies were first formed they made do with cattle trucks, coal lorries or anything they could get their hands on! (*Rex Norton and John Henry Havelock Gray*)

by its nose, with perhaps a magnetic clock-stopper still attached (details of the clock-stopper later), and while in this position the base plate could be removed. Steam tubes could then be inserted in the base of the bomb to melt out the explosives. As the bomb was held upright the melted explosives would run out. The bomb could then be lowered so that the steam pipe was kept close to the diminishing explosives. Once empty of explosives, the fuze and any booby-traps could be destroyed by a small charge of guncotton.[10]

Once a bomb had been made safe, depending on its size, men could either pull it from the excavation using a block and tackle, or larger bombs would be hauled out by a vehicle. On the Continent it is recorded that horses were sometimes used to pull out V-weapon warheads, no doubt because they were more readily available and could cope better than many vehicles in off-road conditions. There were times when bombs that were still dangerous were hauled out manually – the ticking bomb that fell next to St Paul's Cathedral, for example. In these cases only the men needed for the operation would be allowed in the danger area. An unfortunate incident occurred at Castle Street in Swansea, in February 1941. This bomb had some of its explosive steamed out while still in the excavation, to the point where the remainder was well below the fuze chamber. A decision was made to pull the bomb out of the confined space and continue removing the explosives in a safer environment. However, it seems there may have been another fuze that had not yet been exposed. Listening through a stethoscope confirmed there was no ticking and it was therefore thought safe to move. The bomb was connected to a lorry and the officer ordered the men to take cover just in case. He then gave instructions to the lorry driver to take up the strain. While this was going on the rope connected to the bomb suddenly slipped and a sergeant came forward to secure it. On the next pull the bomb exploded. Unfortunately many of the squad had at this point come out of cover to offer their help, unnoticed by the officer. Killed in the blast were Staff Sergeant Thomas Munford, Lance Sergeant Thomas Henderson, Corporals Jack Holder and John Salisbury, Lance Corporal James Johnstone, Sapper William Craig and Driver Roland Simpson. The officer, Lieutenant W. D. G. Rees and the sergeant who re-secured the rope, Sergeant Finney, both sustained shock and punctured eardrums and spent a considerable time in hospital as a result.[11]

The V1 was fitted with a time delay mechanism and mention has already been made of the Germans' fondness for using clockwork delaying devices for their fuzes. When these clockwork fuzes were first discovered in 1940 the British boffins came up with a number of inventions to help counteract them.

Officers would gain some idea of what the fuze was doing by using a doctor's stethoscope to check for ticking. In a quiet environment the ticking

Despite having been issued with heavy lifting equipment, some manhandling of recovered weapons could not be avoided. Rope and lifting strops were an absolutely essential part of a BD team's kit. Here Chief Petty Officer Charles Baldwin (foreground), who was decorated for his part in the recovery of the first magnetic mines at Shoeburyness, is seen in a typical situation for a Navy BD officer. Baldwin was killed shortly after this photo was taken, while attempting to recover another sea mine. (*Author collection*)

could sometimes be heard without one. Many men were tricked by their own watch or even their own heartbeat, as they almost expected to hear ticking. The Germans got wise to this and later developed fuzes where the clockwork mechanism was less audible than before, in the hope that the bomb disposal officers would think the fuzes were not operating. In time, electronic stethoscopes were introduced. This enabled a man some distance away to safely monitor the bomb and let the officer know immediately if the fuze was ticking. These had to be very reliable and work in all conditions. They were fitted with a detachable magnetic element so that they would not fall off the bomb should it come to rest vertically. As a test of the equipment, and the man who was tasked with listening, a watch would be placed on the bomb. They were also able to be fitted with a heat insulating piece, so that a fuze could be listened to while steam was being used to remove the explosives. With V1s it was recommended that two stethoscopes were fitted. One was to be listened to remotely and the other was for the bomb disposal officer

The magnetic clock-stopper 'Kim' was a weighty bit of kit. The lower photograph shows the huge battery boxes necessary for the current. The RAF was a little critical of this equipment as they thought the cells of the battery should be separate units – they believed the equipment did not stand rough handling and leakage of one cell ruined the box. A bungee strap could be used (right) so that the unit could be put up against the bomb without bumping, but it was awkward equipment to use at the bottom of a muddy shaft. Kim was for use on V1s if they were found to be fitted with a type 17 long-delay fuze (see Appendix 1). (*Lt Col Eric Wakeling Ret'd*)

working at the warhead. That way everyone would know what was going on. Even if the bomb disposal officer was killed, others would know at what point in the proceedings the fuze started ticking and how long from that point until it exploded. To help identify the fuze in the first place a torch with a mirror was provided. With this, officers could see round a bomb that might be lodged in a tight spot. It would mean the bomb could remain undisturbed during the initial investigations of whatever fuzes it had.

One of the tools developed to counteract the German clockwork time delay fuzes was a large magnet that would stop the clock's mechanism. It was called 'Kim' so as not to give away its function. Delivery of 100 sets from the General Electric Company took place in October 1940. Kim's heavy weight made it a little difficult to work with. It basically consisted of an air-cored saddle-shaped coil weighing 40lbs, which was placed on the bomb approximately coaxial with the fuze pocket. It was energized by a battery, the current of approximately 200 amperes being switched on for periods of about 3 seconds. Later the magnet was replaced by an iron-cored pot magnet that could be operated continuously.[12] A lightweight magnet could also be operated that could be left on for half an hour before it overheated. The early example became obsolescent, as in practice it was found not to stop all type 17 clockwork fuzes. While removing a bomb from a tight space the clock-stopper sometimes needed to be removed. In the short time it was not connected to the bomb the fuze could restart, particularly if the bomb was being moved, and some bombs did explode for this reason, killing anyone nearby. It happened with a bomb being removed from a house at 590 Romford Road in Manor Park, London. The clock-stopper on this bomb fouled the lifting strop. The clock-stopper was removed just long enough to get the bomb out of the hole. As the bomb was raised it came into view and was seen to be gently swinging. The officer, Captain Blaney, had just stepped forward to steady it when it exploded. Ten men were killed.[13]

Though the V-weapons did not appear until 1944, the lessons learnt from fuzes used by the Germans in aerial bombs of earlier years could be drawn on. The equipment developed for those fuzes had a direct impact on the methods used to render safe V1 fuzes and any potential booby-traps that might be hidden within.

The fuzes in aerial bombs were different from those of V1s in that they were electrically charged and thus armed as they left the aircraft, whereas the V1 used a lanyard to pull out a safety pin on launch. Nevertheless, the two kinds of fuze were of similar size and held in a similar style of fuze pocket that entered the bomb from the side. These pockets could quite easily have contained the same type of booby-trap, a Zus 40, and they also contained picric pellets in rings as did aerial bombs. Other components, such as clockwork mechanisms, were also similar in design to those in normal aerial bombs. Tools that had been developed for aerial bombs were already at hand when the first unexploded V1 was discovered, so were made use of – particularly one used in the 'J' process. To understand how this 'J' (jamming) process worked, one needs to look back to the summer of 1940 and how bomb disposal processes developed.

As mentioned, aerial bomb fuzes could hold the electrical charge in them for some time after they had come to rest, making them still capable of functioning. Early in the war it was found that if the two plungers in the top of the fuze (where the electricity had been 'injected' into the fuze as the bomb left the aircraft), were depressed, then the electricity would safely dissipate and the fuze became harmless.

A conference was held at London's Savoy Hill House on 14 May 1940, to decide on the design of a tool to perform this function and to enquire into its construction. It was decided that Messrs Crabtree & Co. should undertake the work and that Dr Merriman from the Ministry of Supply should go to Walsall, where the firm was based, to assist in the preparation of the design

(Left) Some BD tools in the author's collection including the well-used hammer and chisel. They are made of non-sparking, non-magnetic materials. (Right) A close-up of a Crabtree Discharger. The knob on the side was screwed in to clamp the discharger on to the head of a fuze. The loop on top enabled a piece of string to be tied to it so that the fuze could be pulled out from a safe distance. (*Author collection*)

and drawings.[14] Two days later a pilot model was produced and submitted to the director of scientific research for approval. It says something about the seriousness of the UXB problem at that time that production had commenced by 18 May and deliveries were made during the next ten days. The total number of dischargers ordered was 5,800.[15]

The Crabtree discharger was a small brass tool that could be secured to the top of the fuze, with two prongs that were deployed to depress the two plungers in the fuze. After depressing the plungers about six times the BD officer could be confident that the electricity in both reservoir and firing condensers had drained away and the fuze was rendered inert. It worked well at first but the Germans soon realized what was happening and changed the design of the fuzes – with the new type, when the plungers were depressed the bomb would go off.

As a result, the Crabtree tool immediately became obsolete as a discharger. They still had some use though. The two prongs could be removed so the fuze plungers were untouched and the tool, with its loop on the top, could then have a string tied to it, when an officer wished to pull out the fuze from a safe distance.

Other methods for dissipating fuze electrical charges were investigated, including the idea of injecting a liquid to neutralize the charge. The only openings in the fuze were the two small spaces that the plungers pushed through. These plungers had a seal around them and the liquid would have to be squeezed down the sides. It was very important to control the pressure precisely, as too much would push the plungers down and detonate the bomb. The liquid that was to be used was tested beforehand by the BD officer, to ensure it had the correct conductive properties to successfully work on the fuze. Fortunately it was found that the liquid discharger equipment could also be used on another fuze the Germans introduced, the anti-handling type 50.

If a bomb came to rest embedded in a concrete floor with the fuzes inaccessible, the liquid discharge method became difficult to execute. An alternative was to immunize the fuzes using heat to discharge the fuze's condensers. It was found that the condensers could remain dangerous for 1,000 hours at 60° Fahrenheit but could be discharged after 24 minutes at 150°. Too much heat, however, could distort electrical insulation within a fuze and cause it to detonate the bomb. The RAF experimented using a Steam Jenny, issued to all bomb disposal units for steaming out the explosives from bomb casings. The results of tests indicated that this method of heating fuzes to make them inert should only be used as a last resort as there were too many variables to consider including the strength and direction of the wind, and so on.[16]

A liquid fuze discharger as issued to the RAF in April 1941. It could be connected to a bomb using an extension tube, which enabled fuzes to be treated in any position without actually moving the bomb. This technique in RAF BD was known as 'giving it a drink' with a 'Toby jug' and the immunizing liquid was referred to as 'beer'. Personnel were encouraged to use this terminology as it maintained secrecy, especially in telephone conversations. (*Author collection*)

Another idea for heating fuzes was to paint the fuze and surrounding area of the bomb with black paint and place a heater close by. Again, this did not prove very satisfactory in terms of temperature control. The thermal equipment that did eventually come into service was a sort of kettle affair that had an immersion heater powered by batteries. It connected to the top of the fuze and forced steam past the plungers to the condensers below. This method was not as popular with bomb disposal officers as the liquid discharger.

One piece of kit that did prove a major breakthrough was known as the Stevens Stopper. This was designed to inject a liquid into a clockwork fuze to jam up the works. It could be used instead of a magnetic clock-stopper which was important if a second anti-handling type 50 fuze was also fitted in the same bomb. A magnetic clock-stopper might cause a type 50 to detonate, whereas the Stevens Stopper would not. The S-Set, as it became known, created a vacuum in a fuze that meant the liquid would be drawn into every nook and cranny, as opposed to being pushed in under pressure. In the early days the liquid introduced was a salt or sugar solution. A saturated brine solution had some advantages. If the bomb subsequently had its explosive steamed out, a deposit of salt left in the clockwork as the water evaporated would stop the clock from re-starting, and salt attacked some of the metal components. There were also disadvantages to using a salt solution. It had a low viscosity and, depending on the temperature, might fail to stop the clock or only stop it until the bomb was disturbed. The addition of sugar helped and it was found that the best mixture was 4lbs of domestic sugar to 5 pints of saturated brine. It took up to five minutes for this to completely penetrate the fuze. A downside to using sugar was the fact that, with rationing, men were tempted to steal it. Later the liquid was replaced with a fast setting resin, like that used by dentists for taking casts of teeth. This obtained similar, if not better, results.

There was a shortage of clocks to experiment on, which hampered the work of the scientists. Nevertheless, trials took place using hot liquid, the theory being that as it cooled inside the fuze viscosity increased. This was better for stopping clockwork mechanisms.

With the liquid discharger, using the Stevens Stopper, a hole had to be drilled in the top of the clockwork fuze. If another fuze sensitive to vibration was fitted in the same bomb then the operation could be a very delicate one, requiring a steady nerve and a sharp drill bit. An electric drill was not an option. The S-Set continued to be developed with the full involvement of Major John Hudson. It was to be used to disable more than just the clockwork mechanisms. The Zus 40 spring-loaded booby-trap was targeted as well and ways were found to inject the liquid so that it would fill the booby-trap and stop it from functioning. This was far from easy – problems with bubbles forming inside were one issue the back-room boys had to address.[17]

Obviously Major Hudson was very familiar with the S-Set by the time the V-weapon attacks started and, fitted with a suitable adapter, it was used on the fuzes in those V1s that failed to explode.

Experiments took place to see if a universal liquid could be used to do the lot – dissipate the electrical charge, jam up mechanical and chemical time delays, as well as rendering safe booby-traps.[18] If an all-in-one mixture could be found then less equipment would need to be carried by the BD units. The one thing that most German fuzes had in common was that they shared the same design of gaine – the small cylinder attached to the bottom of the fuze that was filled with explosive. As autumn 1944 approached most gaines found in German bombs were composed of a lead azide/lead styphnate mixture. When this was loosened it could be neutralized by thiobenzoic acid, which had the benefit of also being non-conducting and therefore would not short out electrical connections with the possibility of detonating the bomb. However, it was found that with a Stevens Stopper not enough of this acid could be introduced to render inert the material that was compressed in the gaine. In September, the search for a universal liquid was abandoned.[19]

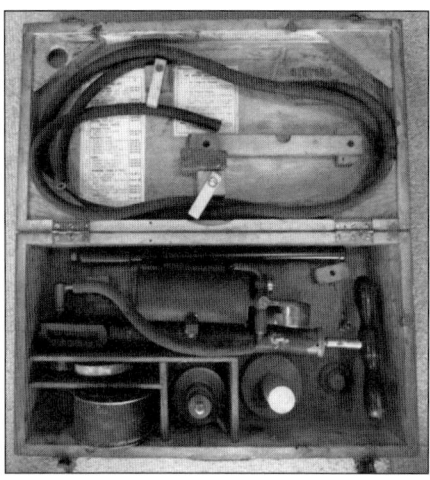 

A Stevens Stopper in the author's collection. Invented by Wing Commander J. C. Stevens of RAF Bomb Disposal (pictured right) it was designed to 'jam up' the works in German clockwork fuzes. For various reasons it did not see wide use until 1944, though it had been successfully tested in 1942 by John Hudson. He was to use this equipment again while working on unexploded V1s. (*Author collection*)

Other experiments examined whether different liquids could be used in the same equipment, but this was simply not practical. Residue left in equipment could have a detrimental effect on other types of solutions used. The authorities realized they were trying to create a jack of all trades but ending up with a master of none.

The Germans came up with another fiendishly designed fuze – the type Y. This was another one very sensitive to vibration. The first type Y successfully recovered was naturally subject to intense investigation. This particular example had failed because of a break in its initiating circuit. The S-Set was used on this fuze, though it was already safe. A barium chloride solution was introduced in order to obtain an X-ray showing the penetration of the liquid.[20] The fuze was filled with liquid under different pressures and with the fuze at different angles. The experiments looked promising, but ultimately the best method found to render these fuzes safe was to make a clay dam around the top of the fuze and fill it with liquid oxygen. This would cool the fuze's battery to the point where it no longer functioned (see photo of Major Hudson in Chapter 1). The opposite method was also tried – heating the battery. For this it was necessary to gain access to the hermetically sealed fuze pocket where the batteries were located and it had to be done without causing vibration or heating the top of the fuze above 70° Centigrade. The ingenious way of breaking into the fuze pocket was to use a corrosive chemical reaction. An airtight container was clamped over the fuze head and a glass tube housing a copper electrode fitted into it. Hydrochloric acid used as electrolyte was contained in two bottles, one connected to the glass tube and the other to the container, with circulation maintained by applying a vacuum first to one bottle, then to the other. Once the seal to the fuze pocket had been broken a 75W lightbulb was directed on to the fuze head from a highly polished reflector, fitted in a conical steel case. The temperature was kept at 60° Centigrade by alternately switching the bulb on and off. In this way the fuze's battery was rendered inert in three hours. However, this method of dealing with Y fuzes was not taken up, the freezing method being preferred.

New procedures and equipment always had to be checked to ensure they had no adverse effects on other parts of the disposal procedure. For example, the resin used to jam up mechanisms might melt with the heat used in steaming out explosives, to the point where the fuze might again become dangerous.[21]

Some more brutal ways of stopping the clockwork were also investigated. Before his death in 1941, the Earl of Suffolk had conducted research into firing a bullet into the mechanism[22] – another method never adopted. A more measured approach looked at using a drill.

The firing pin in the type 17 clockwork fuze was released by a slotted disc that was turned slowly by the clockwork mechanism. The amount of time

Two versions of equipment used to freeze the batteries in fuzes, thereby rendering them inert. (Left) One was called 'F' Equipment, used liquid oxygen and was preferred, as it worked much quicker. (Right) 'D' Equipment used dry ice when liquid oxygen was not readily available. To make the dry ice a CO_2 fire extinguisher was let off into a kit bag. This was then tipped into a container with methylated spirits to form a mushy ice. The ice was scooped into the drum secured over the fuze and the top screwed down to compress it on to the fuze. (*Author collection, Lt Col Eric Wakeling Ret'd*)

On the same day that the first unexploded V1 crashed in the UK, Brigadier General H. B. Sayler, the US Army's Chief Ordnance Officer European Theatre, wrote to Wing Commander Stevens, of Stevens Stopper fame, sending him drawings and photos of the 'Flit Gun'. The covering letter said, 'While an American development, the "Flit Gun", as it is termed, is fundamentally an adaptation of a British principle. It is my hope that this equipment will prove of tremendous value to all operating Bomb Disposal personnel in the Allied Armies.' A few days later it was tried out on the second unexploded V1 (see Chapter 1). The Flit Gun could make a connection into a fuze via a self-tapping needle, so that a thermosetting compound could be injected. The needle would not leak despite the compound being pushed through it under great pressure. It was called a Flit Gun because it looked like a popular insecticide tool of that name. (*Author collection*)

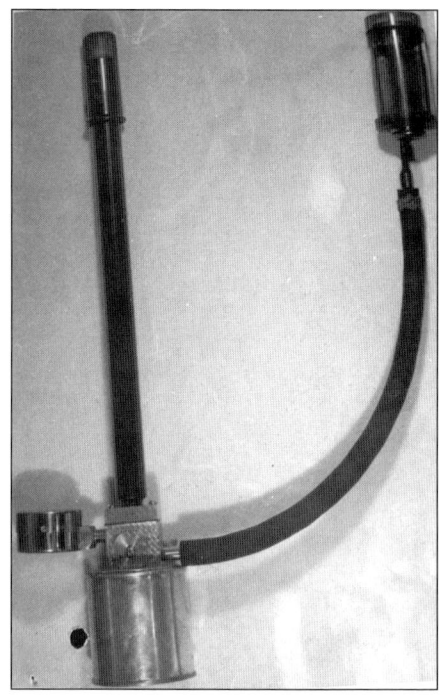

delay was set by adjusting the angle through which the disc had to rotate and this was done by means of a small stud passing through the clock casing. This stud seemed to offer an opportunity but drilling right into the clock mechanism was considered too dangerous. It was thought that the only reasonably safe way would be to drill a hole through the condenser used for initiating the clock fuze, exposing the setting stud, then to drive the stud by impact forward against the slotted disc, preventing the clock from rotating. A device was constructed that clamped to the head of the fuze and when switched on automatically drilled a 3/8in. hole through the fuze. At a prescribed depth the drive to the drill was switched off, but the feed continued and pushed the drill forward as a punch to force the setting stud in against the timing disc. This method, though successful, was considered a little too complicated and was never adopted for use in the field.[23]

Removing a fuze from a bomb could at times be very difficult. As bombs crashed to earth at some speed they could distort or sustain damage in the area around the fuze's locking ring. A number of spanners, or fuze keys, were provided for the use of BD officers, although a hammer and chisel were often resorted to. As it was not always healthy to be up close when the fuze was withdrawn from its pocket, a number of devices were also developed

A Quilter Key in the author's collection. This was a heavy duty spanner with a ratchet, used for removing the locking rings securing German bomb fuzes in their pockets. Designed by Sir Raymond Quilter, it could be attached very securely and the handle extended to give extra leverage. (*Author collection*)

to make this part of the process safer. The Crabtree Discharger has already been mentioned as the most basic way of pulling a fuze – using a long length of string tied to it. This was not ideal as the fuze would be pulled with a jerk rather than a delicate steady pull. The first mechanized tool that was successfully developed for the job was the Fuze Extractor Mk1 – or 'Freddy' as it soon became known.

The Freddy was basically a pneumatic jack powered by the carbon dioxide cartridges that were used in home-made soda water makers. When the cartridge was pierced the carbon dioxide gas would fill the cylinder of the jack and push out a piston, which in turn raised a rod that was attached to the fuze. Some slack was built into the design, so that once the cartridge was pierced there was a time lag before the fuze actually moved. This gave the bomb disposal officer time to get to cover.[24] The whole process took about eight minutes as the fuze was withdrawn very slowly, about one inch per minute. The equipment was not that easy to use in the confined space at the bottom of a shaft as it consisted of a frame that had lots of adjustment to it, to enable it to be fitted to different size bombs.

A slightly different approach was the Merrylees Fuze Extractor, designed by Lieutenant Colonel Merrylees. This was quite simple in design. A tube was

The 'Freddy' fuze extractor provided a way of gently pulling a fuze out of a bomb while remaining out of harm's way should it explode. It was designed by Flight Lieutenant Eric Moxey, who had been involved with bomb disposal from March 1940. He went to France to try to obtain examples of German fuzes but was forced to evacuate via Dunkirk. He was killed on 27 August 1940, while working on a bomb at Biggin Hill aerodrome. (*Author collection*)

screwed into the same thread that the fuze locking ring had used. A drum on the top revolved as cord was pulled and this movement was transmitted to a Crabtree-type clamp fixed to the top of the fuze. A BD officer could pull the cord from cover in a smoother, more controlled manner than just jerking a piece of string tied directly to the fuze. The officer could also tell if the fuze jammed while being withdrawn, unlike using the Freddy where a jammed fuze would not be known about by the officer until he actually returned to the bomb.[25] Lieutenant Colonel Merrylees, incidentally, was very interested in the subject of pendulum dowsing and is said to have successfully used this method to find unexploded bombs.[26]

Not all the unexploded bombs that the Allies had to deal with were German. British and American bombs were found that had either been dropped accidentally, were in the wreckage of crashed aircraft, or were found in locations previously in enemy hands that had been subject to bombing, such as the V-weapon launch sites. The Allies used fuzes (or pistols to use the correct terminology) that were screwed into the rear of the bombs and were armed by a small propeller that rotated as the bomb dropped through the air. The British type 37 pistol was particularly dangerous. It contained an anti-withdrawal device – a striker that would be released as the pistol was removed. In order to extract the pistol safely, before the striker had a chance to hit the detonator, it would have to be done very quickly – in about three milliseconds![27] At the least, the withdrawal should be quick enough for the impact of the striker on the detonator to be so reduced that the detonator would fail to fire.

The French claimed that the Municipal Laboratories of Paris had found a way of doing this with the British type 37 pistol, using what was known as a 'rocket wrench'.[28] These wrenches had been used by the Germans as well as the Allies. They basically consisted of a metal catherine wheel that clamped on to the pistol and two 'jets' that when fired spun the fuze out of the bomb at an incredible speed. The American John Feldman, who was present at the disposal of the first unexploded V1, developed one that used half-inch calibre machine-gun cartridges as the jets.[29] The French and German designs used black powder pellets. After the British performed experiments on these rocket wrenches in 1945 it was recommended that they were not suitable for use against the pistols where the risk of detonation was unacceptable. Again, some slack was designed into them so that they were spinning up to speed before starting to unscrew the pistol. This kind of tool was used in the Pacific theatre on Allied and Japanese bombs and continued to be developed after the war. The rocket wrench was a natural progression from the manually operated wrenches that used a rotating drum with a length of cord, as in the Merrylees, or a length of wire as in the later Mk 7 Fuze extractor.

The Merrylees fuze extractor enabled controlled manual extraction of a fuze from a safe distance. It could be adapted for use with Allied unexploded bombs as well as German ones. (*Lt Col Eric Wakeling Ret'd*)

One of the Americans who was a pioneer in the development of bomb disposal equipment in the Second World War was a man by the name of Robert W. Eigell. In the summer of 1941, before America was officially in the war, Eigell was in the US Navy and along with nine other men he visited the UK to learn how the British were dealing with the threat of sea mines and unexploded bombs.[30] This he did not only through classroom training sessions with the Royal Navy, but also alongside them experiencing 'hands on' disposal of live mines. He returned to the US on 4 December 1941.

Just a few days later the attack on Pearl Harbor resulted in Eigell working on a Japanese unexploded bomb that had lodged itself in the USS *West*

A modified Merrylees fuze extractor being used to remove the pistol from the tail end of an American unexploded bomb. (*Author collection*)

Virginia. To remove the tail fuze he fitted a ratchet type spanner to which he connected two lines. From a safe distance he pulled the line that pulled the spanner in the direction that unscrewed the fuze. Then he returned the spanner to its original position by pulling it back with the other line. In this way he could record the number of pulls required to unscrew the fuze, which would prove useful when it came to the next Japanese fuze to be removed. From this experience Eigell developed the 'Eigell wrench', which the British adopted but misnamed the Igol wrench, (some among the British bomb disposal community say IGOL stands for 'I go on living').

One of the more complicated pieces of machinery developed in the war and still in use is the trepanner. The purpose of the trepanner was to drill an access hole in the side of a bomb, so that the explosive could be removed by the process of steaming out. Although the base plate, through which the bomb had been originally filled with explosive could often be removed, this could disturb a sensitive fuze.

A wartime example of a rocket wrench, with its jaws clamped on to a bomb's pistol. (*Author collection*)

The trepanning machine, known as a 'steam-sterilizer', was a little temperamental in the early days. Often the officer was required to return to the bomb to fix it. The machine used steam to drive a cutting head that drilled through the casing. It would then feed the steam hose into the bomb and emulsify the explosives, which would run out of the newly cut hole. The steam needed for powering the trepanner and for the steaming-out process was produced by a very Victorian looking piece of kit – the Merryweather Steam Boiler. This would be positioned at a safe distance from the bomb that was being worked on.

The trepanner went through several changes of design and an electrically powered one was developed that caused so little vibration while in operation that a coin could be stood on edge on the bomb casing without it falling over. Tests were done and a bomb casing measuring 5/16 in. thick could be cut through in less than 20 minutes, resulting in a hole some 3ins in diameter.

The Navy had its own version of the trepanner for making holes in mines. Developed in conjunction with the National Physical Laboratory at Teddington, theirs was designed to cut holes large enough to be able to remove component parts without using the access routes originally built into them by the Germans. This would hopefully avoid going through the obvious locations where a booby-trap could be fitted. The Navy's machine was not powered by steam, it was pneumatic. Of course, like most BD equipment, these trepanner machines had to be capable of operating in all sorts of conditions. In 1941 a test was made whereby one was immersed completely in a bath of water except for its electrical plug. It was then switched on and in 30 minutes had successfully cut through a steel plate, taking just a few minutes longer than if it had been cutting in air. On examination afterwards it was found that a little bit of water had penetrated the machine, but the test proved that it could cope with some pretty extreme conditions.[31]

Another type of trepanning that the Americans investigated was the use of corrosive acid to cut through steel or aluminium bomb and mine casings.

The acid would be sprayed against the bomb casing with what was known as a 'gravity fed polystyrene spray projector'. This would avoid both the possible vibration from a mechanically operated cutting machine, and any need to move the bomb. A certain amount of practice on the part of the operator was necessary for them to become proficient.[32] The acid method was used on some of the unexploded V1s (see Chapter 1), but a disadvantage was that it took quite a long time.[33] A 35 per cent solution of nitric acid applied to steel bomb cases in a fine spray could cut a hole about 2 x 1 ins through the case, at a rate of about 0.25in. per hour, at a temperature of 60° Fahrenheit, and had no undesirable effects upon coming into contact with explosive. The time increased with lower temperatures in winter. Also the fumes produced were rather unpleasant, particularly in the confined spaces where bombs often had to be worked on.

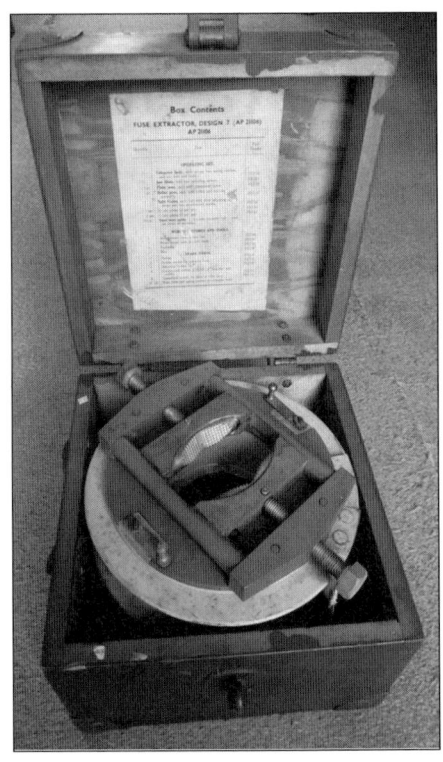

A Mk 7 fuze extractor in the author's collection. It was designed to remove Japanese nose and tail fuzes by unscrewing them remotely. (*Author collection*)

The squad would have to dispose of any explosive removed from bomb cases and the recognized way of dealing with it was to burn it. When explosives are not in the confined space of a bomb case the material is much safer. It is very similar to what happens when children empty out a firework and then put a match to the loose contents – they burn fiercely but without the wrapper they do not go bang (or they shouldn't). However, there is always the exception and explosives in bombs (or fireworks) should always be treated with the greatest respect.

The need for care while burning explosives is illustrated by an accident that claimed both the eyesight and hearing, totally and permanently, of one young man in RAF Bomb Disposal. His name was Wally Thomas and he was attached to 6220 BD Flight, based at North Weald in Essex.[34] His squad was called to the USAAF base at Earls Colne over Easter 1944 because a 1,000kg German bomb had penetrated the tarmac to a depth of 16ft, having

On the left is a trepanner for cutting a hole in a bomb casing, with steam to drive the cutting head. The steam is then used to melt the explosives out of the bomb. The trepanner on the right is powered by an electric motor. (*Author collection*)

failed to explode. It took two days to reach the bomb. The officer immunized the fuze, then a jerk test was done. Thomas was involved in removing the filling.[35] This bomb had a solid filling in the nose and a powdered filling in the rear. The powder was dug out with a wooden spade and put into sacks, and then the solid explosives were steamed out using a steam jenny that Wally Thomas described as looking like 'a broken-down rickshaw crossbred with a mangle'. This too was put in sacks. A suitable place to burn the explosive materials was found nearby – a shallow bomb crater. The men flattened the bottom before starting to burn the explosives.

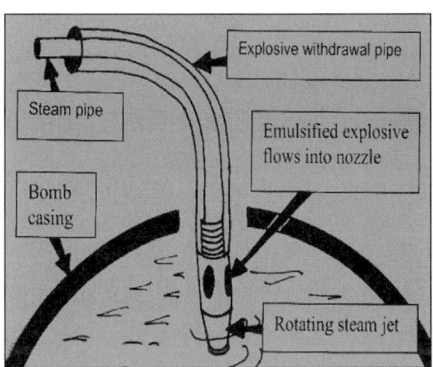

The explosive being steamed out would form a big puddle next to the bomb if it was allowed to run out of its own accord. Pictured above is a diagram that shows how the steam hose was also used to take the emulsified explosives away. Sometimes this hose would block and it was necessary to run it through a drum of hot water to loosen the re-solidified explosive. (*Author collection*)

As the light was by now failing it was decided to finish the job the next morning, Easter Monday. Wally fell asleep that night listening to rain on the roof of his hut: it was the last time he would hear that sound.

By morning the rain had stopped and the burning recommenced. Paraffin-

Explosives removed from bombs were disposed of by burning. With mines the explosive was sometimes burnt while still in the mine's casing. The mine would usually detonate at some point, but if some material had been destroyed there would be a smaller explosion. The mine being burnt in this photo was recovered from the River Thames, 60 yards from the loading wharf of Tate & Lyle in London's docklands. It was quite a recovery job as it had come to rest 15ft below the riverbed. (*Author collection*)

soaked rags in chains were laid to the explosives, then set on fire. Time went on and the powdered explosive had nearly all been burnt. Wally was left by himself by the crater for a while, as his comrade went with a wheelbarrow to get some more solid explosive. While he was away Wally threw some loose pieces of explosive into the fire. He then found some powdered filling that had been missed. He shovelled it up with a wooden spade and turned towards the crater. As he did so there was an explosion. Wally Thomas's life changed in that moment. He lost his sight and hearing, spent a great deal of time in hospital and was subjected to numerous operations. In 1959 he wrote a book, *Life in My Hands*, describing the ordeal he had been through and how he coped with it. No one can be sure why the explosives being burnt detonated on that occasion.[36]

One area of equipment not so far covered in this chapter is that of the men's personal attire. The British forces are known for their strict dress code and 'spit-and-polish' regulations. Accordingly, the men who were first involved in bomb disposal were expected to dress as they would for most other types of operations that their particular service was involved in. Little thought was given to the peculiarities of the job in hand.

Also in the early days, men were expected to dig shafts down to the bomb while still wearing their tin helmets. This was not practical at all – the helmets began to come off as the men repeatedly stooped over. They started to adapt their uniform accordingly, with one squad ending up looking like pixies in bowler hats with the rims cut off, with cut-outs for their ears. They pulled this unusual headgear right down on their heads and it not only offered some protection against loose earth and pebbles that might fall from the sides of the shaft, but also helped a little to keep their hair free of dirt.[37]

(Left) Members of 725 BD section in 1940 wearing regulation tin helmets, having just recovered a bomb from a shaft at Oxshott Woods, Surrey. (Right) An unknown London-based BD section later in the war with a more ragged appearance. Note the holes in their uniforms and a hat that looks as if it might be of German origin. (*Author collection*)

Dirt and water were not a superficial problem. The men often had to put clothes back on that were still wet from the previous day. Maybe that is not so different to the infantry in the field, but the BD men's job of digging out a bomb in wet conditions could at times take weeks or even months. Being covered in wet mud daily could be very disheartening, though Captain Hunt (who worked on the unexploded V1 in a farm cesspit, described earlier) recalled a lighter moment. He was returning to his HQ having just helped some Civil Defence workers extinguish incendiary bombs, when he noticed a smell of burning and thought his car had problems. He found that the smoke that had appeared at first to be coming from under the dashboard was in fact from his boots, which had picked up some of the phosphorus material. As he got out of the car they burst into flames and he had to kick them off fast![38] This phenomenon was also reported by others who worked on phosphorus bombs and, after a number of fires broke out, orders were issued to make sure men cleaned their boots thoroughly on return to barracks. It was found that tiny particles of phosphorus were igniting, as the mud on boots dried out in the night.

The Navy's bomb disposal personnel obviously spent a good deal of their time in wet conditions, often required to wade into the sea or trudge through tidal areas. They had waders and wet weather gear, but if conditions were favourable, such as a deserted beach on a sunny day, more than one officer was known to have gone naked into the surf to recover a rogue mine.

In 1940, the men of the Royal Engineers Bomb Disposal had conferred upon them by the then Queen Mother a battle honour to be sewn on their sleeve.[39] Queen Mary's involvement in the BD badge came about because of her evident interest. She paid Royal Engineers BD units frequent visits from her home at Badminton House, in Gloucestershire, and was photographed with one unit that had recovered a bomb that fell locally in Badminton village, on 9 July 1940. The badge the Royal Engineers BD men were issued with took the form of a yellow bomb with two blue rings round it, on a red background. Though this would immediately show the ARP and Civil Defence personnel the men's credentials for being in a roped-off area, their transport was also easily identified. The BD units' vehicles would have fenders painted bright red, enabling them to be waved through without delay by controlling authorities.

It was not long before the RAF adopted a badge worn in a similar way.[40] In January 1941 the king gave his approval to their badge, which consisted of a bomb surrounded by laurel leaves and with the initials 'BD'.

At the same time the Royal Navy was looking into using a similar design to the Royal Engineers, but with a red bomb on a black background, with some other colour variations depending on the different uniforms worn. Surprisingly, the Director of Torpedoes and Mining (DTM) was not too keen on the idea of these badges for Rendering Mines Safe personnel. He felt that the issue of a 'special badge' was not justified and also thought that just because certain Naval Bomb Disposal Officers had been trained in the work and might be called upon to assist in an emergency, they did not need a badge.[41]

Though an example of the Royal Navy's sleeve badge is held at the National Archives, a period photograph of one actually being worn proves elusive.[42] The DTM thought they were simply not suitable for officers' uniforms. Often the Navy BD officer would 'borrow' men from the lower ranks, from whatever naval unit was closest, in order to complete a job. They were not trained in bomb disposal and did not really need to be for the work they were given in most circumstances. These men would most certainly not be issued with any form of BD insignia.

The Royal Engineers styled insignia was adopted by many other nations with slight modifications and is still in use with quite a few forces around the world today. The American Army in the Second World War wore a red bomb with black background on the opposite sleeve to the British. Sometimes they would have it sewn on their jacket breast pocket for more prominence (see photo of Lieutenant George W. Collins in Chapter 4).

The way today's bomb disposal officer dresses, in a protective suit, goes to show how far materials and ideas have developed since the 1940s. Very little protective clothing was provided in the Second World War and, of course, with a large bomb the clothing would make no difference if you were close by when it exploded. However, had the same style of suits and helmets been available for men back then, some lives no doubt would have been saved on minefields and at incidents dealing with smaller items of ordnance. Probably the only clothing of the time that could accurately be described as 'protective' was gloves to wear when handling explosive. To the naked skin explosives could be very harmful. A number of BD men were forced to give up the work after contracting chemical dermatitis. Able Seaman John Tuckwell served for 18 years with the Navy and in later years was severely crippled by this complaint.[43] He had been involved in the steaming out of explosives from numerous mines and was awarded the George Cross for his work with Sub Lieutenant John Miller, rendering safe a mine in the mud of Barking Creek in East London.

Other protective clothing issued in the war was intended more to protect against the elements than the bombs. Wing Commander J. C. Stevens made

Photographed in the summer of 1940, these members of an RMS (Rendering Mines Safe) team are removing explosives from a mine. The material is being handled with gloved hands as it was very harmful to the skin. (*Author collection*)

sure that each man involved in RAF bomb disposal was entitled to be issued with an oilskin coat, jacket and leggings. He also complained that officers in his section suffered considerable wear and tear to their service dress, as they were frequently called out at short notice to take charge of field operations involving bombs in crashed aircraft or buried in the ground.

During the Second World War improvisation always played a large part in equipping the BD men for the job in hand. If new types of unexploded ordnance were found, then the officer would have to hope his existing equipment could be used or adapted. If not, they might need to devise a new tool. Such was the case when Squadron Leader Scamell dealt with his first V2 rocket warhead and had to have a local blacksmith make a spanner to fit it.[44] Similarly, with the recovery of the first magnetic mine from the mudflats at Shoeburyness in November 1939, Lieutenant Commander Roger Lewis used a page from a signal pad to take an impression of the securing ring, so that a 4-pin spanner could be made out of brass overnight.[45]

Things did not always work out so well. The first death of a US Navy BD man in the UK occurred while he was assisting a Royal Navy mine disposal officer with dismantling a mine at Corton Sands, near Great Yarmouth, on 11 June 1942. The American, Ensign John Howard, was last seen shaping a piece of wood to make a wedge as the Royal Navy officer, Lieutenant Commander Roy Edwards, attempted to prise open a cover plate. The mine suddenly exploded killing the two men instantly. Later it was confirmed that the mine contained a booby-trap.[46] In 2002 a memorial plaque to the men was installed near the spot.

Disarming unexploded bombs in the Second World War was an international combined effort that included the lower ranks who exposed the weapons, the officers who disarmed them and the back-room boys who

Dutch BD personnel in March 1982 with a V1 warhead that had split open on impact in the town of Markelo. The respirators and gloves were worn for protection against the toxic explosives that had just been removed and bagged up. (*Stichting Geschiedkundige Verzameling EOD, Netherlands*)

developed the equipment needed. Despite the specialized equipment, trying to produce a machine that could duplicate the thought processes that a human being calls 'judgement' is something that was, and still is, some way off. Our senses have evolved over millions of years and an officer could see or feel what was happening in a way that a machine could not. When a component was removed for the first time, for example, a machine would not have been able to feel if there was 'too much' resistance, if there was perhaps something hidden, connected to the other end. For all the technology available, it was often better for a man to work up close to the bomb to obtain a successful outcome. This certainly proved to be the case with the V-weapons.

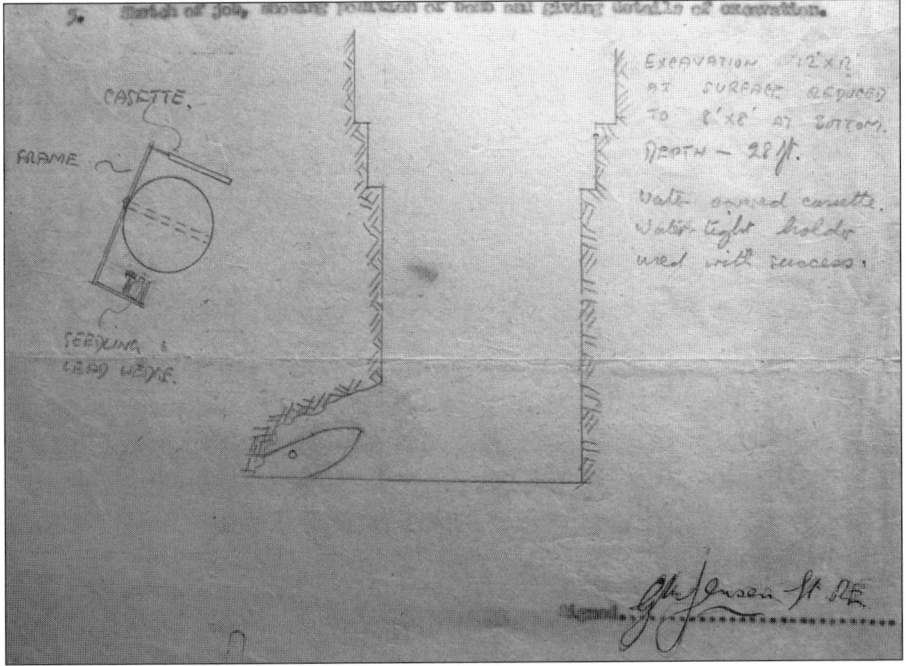

Equipment for X-raying bombs was not issued to BD squads as a matter of course but was held at central points, to be provided as necessity dictated. As mentioned, X-rays were taken to check the fuzes of unexploded V1s, but they had also been used earlier in the war. An example is shown by the diagram above made by Lieutenant G. M. Jensen. It is of a 1,000kg bomb he worked on with Lieutenant A. C. Thomas at Preston, Hull on 31 March 1944. From his notes one can see that the bomb was located at the bottom of a shaft 28ft deep and that the cassette containing the film for the X-ray was in fact covered by water. Both Jensen and Thomas were awarded the George Medal for their work on disposing of butterfly bombs that had fallen in Hull. One that Jensen successfully dealt with was found completely submerged in faecal matter at the bottom of a seven-way sewer junction. The material was removed from the bomb using a spoon he 'borrowed' from a Salvation Army canteen! (*Author collection*)

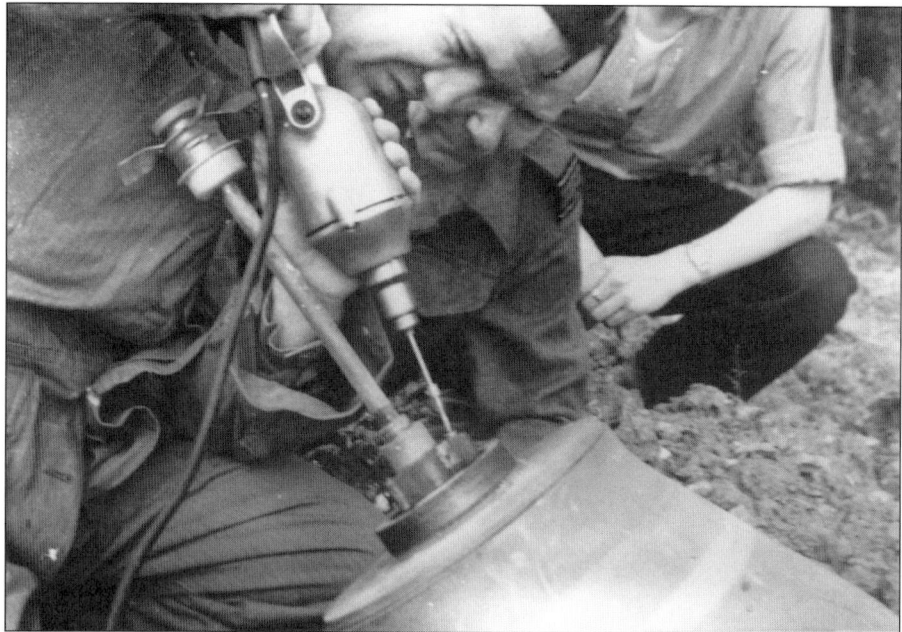

A joint effort of US Navy, US Army Air Force and Royal Air Force, working on a new method of inducing jamming fluid into an American bomb that had been dropped in a cornfield at Hallingbury, Essex, in August 1944. (*Author collection*)

Chapter 6

Post-War Discoveries

For a few years after the war there were actually a fair number of V-weapons to be cleared on the Continent. It has been estimated that over 1,200 V1s crashed prematurely.[1] How many were actually UXBs, defuzed and taken away is subject to debate.

As the war drew to a close, these crashed V1s continued to cause casualties. The scrap metal 'leftovers' from the conflict became a focus for locals who wanted to recycle parts. The wire that was wound around the large compressed-air bottles in the V1s came in handy. So too did the fuel tanks that could be cut in half to make useful containers. It has also been said that farmers attempted to use the explosives from the warheads as fertilizer for their crops.[2] It is no wonder that accidents occurred!

In the spring of 1945, two boys and a 60-year-old man were killed in the Schalkenmehren area of Germany by an exploding detonator – the explosions

The wreckage of a V1 found at the Nijreesbos launch site, Almelo in the Netherlands, after its liberation. (*Hermann Hinsenveld*)

This V1 warhead was located by the Bolkshoek launch site at Almelo. Notice the two fuze pockets. In the background is the rest of the fuselage with the distinctive compressed-air bottle visible inside. (*Hermann Hinsenveld*)

were heard in Manderscheid, some distance away.³ In an attempt to address the problem in the Eckfeld region residents were asked to check their locality after church service on Sunday and notify the authorities if they found any dangerous ordnance lying about.

There was no systematic search made for unexploded V-weapons; instead, they were dealt with as and when they were found. For example, 6224 BD Flight destroyed three V1 warheads at Kaltenkirchen in Germany on 10 October 1945 and another seven on 15 October. The war diary for 6208 BD Flight RAF records that they made a visit to a bomb dump at Leck in northern Germany on 17 January 1946. They found the dump contained some 228 tons of munitions, plus 90 V1 warheads. On 22 January, demolition of the warheads began at Westre. This went on for some time – it was 18 April before the Leck bomb dump was declared totally cleared.⁴

Other units were also destroying flying bomb warheads at this time: 6224 BD Flight dealt with a quantity of V1 warheads at Krummel, on 29 January

1946. The weather was so cold at some of these demolition sites that as water was blown out of existing bomb craters, it froze before it hit the ground. The journey to the demolition areas and sites of UXBs could be treacherous in the harsh winter weather. Mud bogged down lorries and there were a number of accidents caused by icy roads.[5] Flight Lieutenant Fleming of 6205 BD Flight had had first-hand experience of this during Christmas 1944, when the Ford Utility he was driving skidded off the road and crashed into some cottages.[6]

As winter turned into spring the work of the BD organizations continued. Things were beginning to get a little better organized. On 8 April 1946, Flight Lieutenant A. Fenton and Sergeant Edwards left 6234 Flight HQ to attend a meeting at The Hague, with Captain Smith of the Royal Army Ordnance Corps and Squadron Leader Lewis of the British Military Mission to the Netherlands, concerning unexploded flying bombs reported in the northern areas of Holland. At the meeting it was arranged that a recce would be made the next day of six of the bomb sites. Fenton and Edwards joined Smith

A damaged V1 in woodland in the Tondorf region. The warhead casing has been torn open, exposing the explosives. The side fuzes are clearly still in place. (*Hans Joachim Ulmer*)

and Captain J. Van Sleezen of the Dutch BD organization at Appeldoorn. A number of sites where V1s had been reported were visited and the men met up with Lieutenant O. J. J. H. Aalpol of the Dutch BD. Having confirmed the presence of UXBs, the men made a plan of action for dealing with all of them and a report was subsequently sent to AIR HQ (Admin) British Air Forces of Occupation, so that they were aware of what was going on.[7]

On 4 September 1946, 6201 BD Flight was asked by the military government in Germany to inspect unexploded V1s at Aachen. No precise information was available and, as on a previous occasion, a recce party walked in vain through miles of woodland and meadows looking for reported V1s. The military government was asked to contact the Flight again once they had definitely pinpointed the UX V1s in the Tondorf area. On 23 September they came back with some better information on the locations. A reconnaissance party was shown two of them – one at Nettersheim and another at Blankenheim. These were demolished on the 30th, the day after they had been discovered. The Flight had found four more V1 warheads intact at Lommersdorf and one at Rippsdorf and demolished these on 28 September. The area around Tondorf over the next few days was reconnoitred in search of further V1s.[8]

It was not just the Royal Air Force that was recovering crashed V1s. The Royal Army Ordnance Corps had 'Enemy Ammunition, Depot Control Units' (EADCU) that worked with German personnel. The ordnance they collected was either dumped in the sea, or taken to a suitable location and then detonated en masse. It would appear that they did not attempt to defuze V1s or move them, but simply blew them up wherever they were found, even if this meant nearby properties were damaged. In cases where a V1 was badly smashed up and the explosives exposed, they would attempt to burn out the explosive. The heat generated would normally explode the detonators after a while and in turn whatever explosive had not already been consumed by fire.[9]

German bomb disposal personnel were also involved with the clear-up. In 1946 one of their experts named Van der Velde was killed, as a V1 fuze he was handling went off.[10]

In 1951 a German magazine called *Spiegel* published an article (26/1951) about the unexploded flying bombs in the Eifel region. A German bomb disposal expert, Eric Kayser, who had 23 years experience working with explosives, described the problems faced. According to him, the EADCU destroyed a total of 24 dud V1s, but because their practice was to blow them up, locals were reluctant to report on where the bombs were sited. If they were hidden in the woods then those who knew would warn local children, but they would not tell the authorities as they didn't want to run the risk of damage to their houses.

Kayser had his own ideas on why so many V1s crashed prematurely in the Eifel area. The launch sites had a valley running behind them and he was of the opinion that the turbulent air coming off the hills affected the V1's stability. He also said that some of the warheads had not been completely filled with explosives (the Germans were suffering from a lack of resources by

The wreckage of a V1 that sat in a forest in the Blankenheim area of Germany for seven years before finally being disposed of. (*Hans Joachim Ulmer*)

Walter Mitzke (left) and Joachim Ulmer (right) removing the fuze pocket from a V1, having cut away the surrounding casing. (*Hans Joachim Ulmer*)

the end of the war), and this he thought influenced the V1's balance. Kayser found that in some cases, instead of 850kg of explosives, there was only about 500kg.

By 1949 the Germans had taken more responsibility for clearing the leftover unexploded bombs in their country. At that time 13 such V1s were known about. Using the EADCU's methods the Germans blew up the first one successfully. The next one, however, was close to houses in Blankenheim. It was decided to try to get drawings of the internal arrangements of the firing systems, so an attempt could be made at rendering this V1's warhead safe.[11]

Rolf Bayer, one of the specialists who had worked at Peenemünde, was located now working for the British. He knew how the detonator operated and provided a drawing. Around the same time another German bomb disposal man, Hubert Schmitz from Dusseldorf, worked on a crashed V1 that had a warhead with a casing made of plywood. On 16 April 1951, he successfully recovered the fuzes and it was then obvious to the men that if they pushed a piece of wire, similar to a paperclip, into the fuze, they could block the spring-loaded mechanism from working. The wire couldn't be too long or it would be in danger of detonating the fuze if it was forced. A bit of sticky tape could prevent the wire from falling out, particularly if the fuze had come to rest upside-down.[12]

The disposal of V1s continued and at least now the men were now armed with a little more knowledge. In May of 1951, bomb disposal personnel Walter Mitzke and Joachim Ulmer dealt with a V1 that had been sitting in the forest in the Blankenheim area (followed by eight others also found in the vicinity). A few weeks later two more V1s were disarmed in Bronsfeld and Milzenhauschen districts by Schmitz and Kayser.[13]

While drainage work was being carried out at Piershil in the Netherlands in 1967, a large object was uncovered. The landowner immediately realized what it was, as during the war the area had been under the flight path of V1s launched from the Pernis area, south-west of Rotterdam, towards Antwerp. It was known that one flying bomb had crashed unexploded in the winter of 1944/45, on farmland near Piershil's creek, and had been rendered safe by the Germans. The wreckage found in 1967 was dealt with by a Netherlands bomb disposal team.[14]

Probably the last person to be killed as a result of a V1 was in 1977. According to the German e-newspaper (*www.volksfreund.de*), 69-year-old

Walter Mitzke defuzing a V1. (*Hans Joachim Ulmer*)

In the summer of 1975 this V1 warhead and fuel tank were found in the Netherlands' Flevo Polder. (*Stichting Geschiedkundige Verzameling EOD, Netherlands*)

Another V1 found in the Netherlands in 1975 was discovered in the Puttershoek area. Note the fuze that has been broken into pieces. (*Stichting Geschiedkundige Verzameling EOD, Netherlands*)

Post-War Discoveries 175

In 1976 this V1 wreckage was dug out of the ground in the region of Oud-Beijerland by a Dutch bomb disposal unit. The explosion shown is the controlled detonation of the fuze pockets. (*Stichting Geschiedkundige Verzameling EOD, Netherlands*)

In the same year another V1 was found near Oud-Beijerland at Puttershoek. (*Stichting Geschiedkundige Verzameling EOD, Netherlands*)

George Mayer of Lissendorf, while out for a walk, came into contact with a fuze that had most likely been removed from a V1 some years earlier. It exploded, killing him.[15] In the period following the war it was apparently quite often the case that fuzes removed from V1s by BD personnel would be discarded nearby. The priority at that time was to render safe the warhead as a whole. With the passage of time these discarded V1 remains were recognized as a danger.

In October 2000, in the Hänscheid area of Germany, an unexploded V1 made the news when it was discovered during construction work. It was found broken into three main pieces and the warhead was complete with detonators. It was disarmed by BD experts Peter Bens and Herbert Schneider, before the Cologne District President, Franz Joseph Antwerpes, had his photo taken with the wreckage. The warhead was removed and later blown up in a controlled explosion.[16]

In 2001 another V1 was found in the vicinity of Spich in the Troisdorf region (south-east of Cologne) and was rendered safe by a bomb disposal unit.[17] Volker Lessmann worked in the German Explosive Ordnance Disposal organization for 18 years and apart from the Hänscheid V1 the year before, this was the only

In May 1980 this relatively complete V1 was recovered at Spijkenisse Hekelingen in the Netherlands. In the bottom photo the fuel tank is being drained. V1s could use low grade gasoline and early examples held 640 litres (169 US gallons). Later ones were built with the warhead reduced in size a little to increase the fuel capacity and thereby obtain extra range. (*Stichting Geschiedkundige Verzameling EOD, Netherlands*)

one he had come across. In this case the explosives had been removed by the German military in 1945, but the bulk of the V1 had been left in the ground, along with its fuzes. Locals remembered that the fuel was also removed. Volker Lessmann believes that at the end of the war the army had no heavy excavating machinery and it was just too much work to dig out the whole carcase by hand (one of their hand-tools was actually unearthed at the site in 2001).

A V1 recovered from the village of Eefde in the Dutch province of Gelderland in April 1980. The house in the background would no doubt have been destroyed had this one exploded. (*Stichting Geschiedkundige Verzameling EOD, Netherlands*)

A couple of fuze pockets with no. 80 fuzes in place being held by members of a Netherlands BD unit in the 1980s. These were from a broken-up warhead that had a wooden casing. The man in the centre is Antoon Meijers. He spent 32 years in bomb disposal and during that time worked on four V1s – three with wooden cased warheads and one with a steel casing that had split open on impact. (*Stichting Geschiedkundige Verzameling EOD, Netherlands*)

(Top) An excavation of a V1 in June 1980 in the Gorssel region. (Bottom) Another V1 was found in Gorssel in October 1981. Many V1s have been recovered by Dutch EOD units from this area over the years. (*Stichting Geschiedkundige Verzameling EOD, Netherlands*)

March 1982 saw a V1 being dealt with in the Netherlands at Markelo. Two no. 80 fuzes were found. The warhead was split, so the casing was bent open and the toxic explosive removed. This was put in the plastic bags pictured here, with Gabriel Doreleijers and Antoon Meijers. The fuze pockets were later detonated. (*Stichting Geschiedkundige Verzameling EOD, Netherlands*)

A couple of years later, in early 2003, more V1 parts were discovered. This time some ominous looking metal objects were found protruding from the mudflats 600m south-east of the island of Neuwerk, near Cuxhaven. A report was made to the authorities, who initially thought the remains could be sea-mine related. Experts were flown to the site by helicopter and they identified the parts as compressed-air spheres from a V1. No explosives were found, so the spheres were taken first to Neuwerk and then by boat to Cuxhaven. The German press at the time reported that the V1 these parts came from was actually fired from the Nordholz area after the war as a 'test rocket'. Both a Nordholz museum and a museum of the Hamburg Police expressed an interest in obtaining one of the compressed-air containers for their respective collections.[18]

After the war, farmers in the area around the former monastery at Bucholz, near Eckfeld, needed to rid their land of unexploded V1s. They used horses to pull them to a more convenient spot where they could be buried. A witness described how the V1s were dragged into an old bomb crater and then covered over with earth. On 3 August 2004 a careful excavation of the

This V1 was discovered buried in Hekelingen dyke in Spijkenisse, to the south-west of Rotterdam in August 1982. Those involved in disarming it included (L to R) Sergeant Kuiten, Captain Van Maren and Sergeant Major Jan Fiers. (*Stichting Geschiedkundige Verzameling EOD, Netherlands*)

A V1 was discovered in woodland in 1984, in the Spijkenisse area of the Netherlands. It was found that a tree trunk had distorted the warhead's casing. (*Stichting Geschiedkundige Verzameling EOD, Netherlands and Raymond Rutting*)

In October 1992 a V1 was found in the Netherlands at Voorst, north-west of Arnhem. It was taken out to sea off Den Helder by the authorities and destroyed. It is normal practice that after recovering large bombs, the Dutch Duik-en Demonteergroep Koninklijke Marine (an EOD Dive unit), take them to the North Sea for demolition. In the past the Dutch had three separate EOD units – Army, Navy and Air Force. In 2009 they were amalgamated into the Defense EOD Unit within which there are two EOD companies: (land) and (maritime). (*Dutch Institute for Military History collection*)

Herbert Schneider (centre right photo) was part of the bomb disposal team that dealt with a V1 found in the Hänscheid area of Germany in 2000. It still contained its fuzes. (*Volker Lessmann*)

Among the remains of this V1, found near Cologne in 2001, were the fuzes. The hole in the top of the fuze, from where the arming pin was pulled out on launch, is just visible. The fuzes were destroyed in a controlled explosion. (*Volker Lessmann*)

Post-War Discoveries 185

In 2004 near Eckfeld, Germany, excavators were used to raise a V1 fuselage and wing spars from the spot where it had been buried by farmers at the end of WW2. (*Jochen Tarrach*)

Recovered wreckage is inspected. Notice the pulse jet engine standing to the right of the photo. (*Jochen Tarrach*)

site began. Not long into the digging recognizable V1 parts began coming to light – compressed-air bottles, pulse jet engines and other rusty debris. It was estimated that the remains of six to eight V1s were present. Among the rusty remnants were a number of warheads that still contained their fuzes. These were disarmed by the authorities.[19]

In February 2007, during construction work close to Teuge airport in the Dutch province of Gelderland, a mechanical digger unearthed something that looked very much like Second World War ordnance. The authorities were informed and a BD unit went to the scene. The remains were identified as a compressed-air bottle from a V1. The area was searched with a metal detector but only small readings were found. A digger removed some earth and exposed more parts of the weapon, including a 'pipe' but no warhead or fuzes were present. A small amount of explosive was used to open up the compressed-air bottle to ensure that there was no possibility of any residual pressure still inside. These rusty relics were then taken away.

A very similar event took place just a couple of months later, in April 2007, on the West Frisian Island of Schiermonnikoog, when a rusty-looking object was found on one of the beaches. A bomb disposal squad arrived by helicopter and quickly ascertained that the relic was another compressed-air sphere from a V1, but no other parts were present. Again explosives were used to open up the bottle and it was then consigned to scrap. Within hours the beach at this popular tourist destination was reopened to the public.[20]

As well as pieces of V1, occasional V2 finds have been made in recent years. In January 2006, a German bomb disposal unit under the command of Dietmar Schmid excavated a woodland site south of Route L-309 in the Westerwald district, where a V2 had crashed back to earth soon after launch. Many mangled parts of the rocket were recovered including pieces of turbocharger and compressed-air bottles, but no explosives turned up.[21]

During construction work at Prüm in Germany in September 2009, the remains of a V1 were discovered. A bomb disposal team soon found that the warhead was not among the wreckage.

In October 2009, a flying bomb was dug up close to the town of Zilsdorf in Germany, after a member of the public reported the possibility of a V1 being buried in the locality. A bomb disposal team of eight specialists, led by Horst Lenz, began a careful excavation which necessitated the closure of the road between Zilsdorf and Betteldorf. Within a few hours the remains of a V1 were uncovered. The fuel tank was found about half a metre below the surface and about 800kg of explosives were found another metre down. The warhead casing was one of the later versions, put together when raw materials were in short supply, and in this case it had been made from wood. After sitting in the earth for years the wood had rotted away.²²

This was the first V1 Horst Lenz had been required to deal with in 25 years of service. During the excavation it was found that there were no fuzes present. Because the explosive was known to be highly toxic the team wore protective clothing, as well as breathing masks, while they worked. Despite the fact that the explosive, amatol dinitrobenzene, is not water soluble, the soil around it was removed and taken away as a precaution and although museums had expressed an interest in acquiring the V1 as an exhibit, the warhead portion was taken to an ammunition dump near Koblenz and destroyed.

Back in the UK there were few reports of unexploded V-weapons after the war. Jim Jenkinson served with RAF Bomb Disposal throughout the war and in fact spent nine years as a sergeant in 5131 BD Squadron. Soon after, he was posted to 6235, a small unit at Acaster Malbis, an airfield opened as a satellite to Church Fenton, not far from York. He spent the first two weeks sorting out paperwork, as the rest of the squad were clearing a practice bombing range somewhere on the north-east coast – it was actually a relief to him when the first 'call-out' was received. Apparently a flying bomb had been reported at RAF Yeadon (now Leeds/Bradford Airport). His first reaction was 'rubbish', but nevertheless he accompanied his CO to investigate. The CO read up his manual on V1s and, on reporting at Yeadon, was directed to a ditch behind the station stores. Sure enough there sat the carcase of a flying bomb. It had been stripped of everything including the warhead, gyro and motor. When the stores officer asked what they were to do with it, the CO replied curtly, 'Stick it in your scrap yard and don't waste our time in the future!'²³

On 28 July 2008, the BBC reported that an unexploded bomb, thought to be a 'Doodlebug', was unearthed at a building site on London's Isle of Dogs. Police set up a 400m exclusion zone in Docklands. The incident generated a fair bit of media interest but little follow-up to the story can be found, no doubt because the police later reported there were no explosives present.²⁴ The bomb squad apparently found that the item was nothing more than an old metal tube filled with concrete!²⁵ Similarly, on 13 August 2010, there were reports that a large object had been found in the Thames near London Bridge. It had been spotted by the Port of London Authority who were using a new scanner to look at the riverbed. News agencies at the time speculated that it 'could be an unexploded V2', but this did not prove to be the case.²⁶

It is possible that unexploded V1 or V2 warheads could yet come to light in the future, as building or dredging work goes on both in the UK and on the Continent. In 2000, the Environment Agency in East Anglia was dredging the River Stour and found a large piece of V1. Though only a pulse jet engine

Many V1 relics were found near the town of Almelo, Holland, in 2008 by history enthusiast Thomason. The items included a pulse jet engine (pictured), a piston used to launch the V1s, the propeller and part of the air log, and a bulkhead. (*Thomason*)

A V2 rocket engine was discovered by a friend of the author in 2009, while he was out for a bike ride in an Essex wood not far from London's busy M25 motorway. (*Author collection*)

from an exploded example, it shows how relics from the Second World War can remain for many years hidden from sight.[27]

There have been cases of excavations turning up rocket parts in recent years in the strangest of places, for example at Freeman Field in Seymour, Indiana, USA, in the late 1990s. Pits were dug here to look for aircraft parts from the Second World War that were discarded at the end of hostilities. This airfield had been a storage area for trophies brought back from the war. Among the items found buried were apparently some pretty hefty V2 parts.[28]

A little time surfing the Internet can turn up photos taken by metal-detecting enthusiasts on the battlefields of Europe. It is not uncommon for them to dig up dangerous-looking ordnance. Among the photos the odd piece of V1 can be seen, not just small fragments but complete engines, compressed-air spheres, a bulkhead, rudders, etc. Other parts of these weapons amazingly

In March 2012 a heavily damaged combustion chamber from a V2 was recovered close to the sea wall on the east side of Wallasea Island, Essex. A number of V2s are known to have come down in this area. Remote locations are often involved when it comes to finding surviving relics from the war and the fact of it being a heavy chunk of metal, difficult to handle, has prevented souvenir hunters over the years from removing it. This area was to be redeveloped as a nature reserve and parts of it were to be flooded, so the piece of V2 has finally been moved. Thanks to the efforts of the Royal Society for the Protection of Birds and Essex County Council, it will be preserved and put on display to the public. (*John Myers*)

are still to be found lying on the surface in some places in the British and continental countryside. As we have said, there are sure to be some warheads that remain unexploded somewhere, perhaps hidden beneath the waters of the English Channel, the North Sea, or in rivers and marshes.

The reed-covered shoreline near Peenemünde was home to a V2 warhead for many years. It was on 30 March 2011 that fishermen from Kröslin found the well-preserved nose cone standing upright among the reeds.[29] There was no obvious damage to it as a result of impact or from the rocket's destruction in the air. The authorities were informed and they quickly established that the warhead contained only sand and bitumen, as was the case with test rockets. The sand replicated the weight of an explosive filling.

With the help of a buoy-laying vessel, the *Oie*, the nose was recovered and was later donated to the Historical and Technical Museum at Peenemünde. The museum staff believe that this relic may be a part of a test rocket fired on 19 February 1943 from the tip of Usedom. On taking off a fire broke out at the rear of the rocket and instead of travelling the planned 200 miles, it crashed just a short distance away.[30]

Salt marshes in Essex also concealed for decades a remnant of the V-weapons programme. On the same day as the *Daily Mail* reported that a V2 had been found in Harwich harbour (mentioned in Chapter 3), another V2 engine was being recovered 30 miles away at Wallasea Island on the River Crouch. This work took place on 29 March 2012, contractors having discovered the relic while preparing the area for use as a nature reserve. Volunteers from the Royal Society for the Protection of Birds and an Essex County Council field archaeologist, Ellen Heppell, dug out the weighty piece of rocket.[31] Then Elderton and Sons provided the mechanized lifting gear needed to move it over the seawall. One of the volunteers, Ian Moon, commented, 'When I began volunteering for the RSPB I never expected to be asked to plan the recovery of a wartime rocket!' In this instance there was obviously no need for any bomb disposal personnel to be involved. The engine had to be recovered as quickly as possible since the area was about to be re-landscaped. It was then to be donated to a local heritage centre and used for educational purposes.

Chapter 7

Fact or Fiction?

While undertaking research for this book a few 'tales' and eye-witness accounts emerged that are worth repeating – some more credible than others. There may be an element of hearsay or secondhand folklore, but generally these accounts are from people who truly believe they were witness to V-weapons that did not explode. It doesn't help that there are rarely firm dates that could be checked to help verify these stories. Some recollections are from people who were very young at the time and one must allow for the memory to be a little hazy with the passage of time, but still, some of these reminiscences are certainly worthy of further investigation.

One such intriguing account was received in response to enquiries made regarding the V2 that came down at Hutton, contained in a letter from Mr Robin Gaymer of Mountnessing, Essex. Though he couldn't help with the V2 incident, a friend of his gave an account of an unexploded V1 that came down in the locality. It landed in a field to the east of the farmhouse on the right-hand side of Thoby Lane, just past St Annes Road, on the way to Swallow's Cross. Mr Gaymer's friend recalled, 'It came down and demolished a chicken shed in the field and stopped short of a horse trough at the side of the lane leading to the farm. I actually saw the machine as it came in from the east and the engine on it stopped just before it reached Holbrooks [a house on the south side of Thoby Lane]. I believe the machine was taken to Hornchurch aerodrome when it was defused.' This would most likely have been an air-launched missile as it had come from the direction of the North Sea.

One evening during the war the air-raid siren alerted residents of the Sherrington Park area of Ipswich to impending danger and before long a doodlebug was seen. Like the V1 mentioned above, this one was described as approaching from the east. The engine stopped as it passed overhead. Those watching sought the safety of the air-raid shelter, but heard no explosion. Apparently Ipswich's *Evening Star* newspaper reported later that the V1 had come down without exploding, in a field just a mile or two away at Whitton.[1] If so this is unusual, as press reporting of locations where V1s came down was banned, in case it assisted the Germans by giving them useful information

At Thoby Lane, near Mountnessing in Essex, a V1 apparently came down without exploding and slid across this field from left to right, stopping next to the lane to the farmhouse. When photographed in 2010 the farmhouse buildings were obscured behind fir trees to the right of the photograph. (*Author collection*)

The Thoby Lane V1 was thought to have been taken to RAF Hornchurch. Danny Kilgariff (centre) served with 6226 BD Flight and was based at Hornchurch in the mid 1950s. While there he recalled seeing four or five complete V1s on site, one of which is pictured above. Where they came from and their ultimate fate is not known. (*RAF Bomb Disposal Association*)

regarding the accuracy of their attacks. A Civil Defence circular issued by the Home Office on 26 July 1945, stated that there was no objection to publication of full reports and photos of individual flying bomb and rocket incidents, including the exact locality, provided that, since 8 September 1944, the date was not given more exactly than by the month and no maps, statistics, figures for casualties or value of property damaged were given out.

Not all V1s came down in the direction you might expect. Some had been known to turn through 180 degrees and head back out to sea again. A change in direction could be due to a malfunction, perhaps as a result of ack-ack or fighter attack. Fighters could sometimes also flick a V1 with a wing, causing them to veer off and hopefully crash before reaching densely populated areas. One eye-witness reported seeing a V1 travelling west to east over Suffolk, south of Diss. It was in a shallow dive with the engine running erratically. On hitting the ground there was no sound or evidence of an explosion. It was suggested it might have impacted in woodland somewhere west of the A140, between the villages of Thornham Parva and Thornham Magna.[2]

Residents in the area of Normandy Drive, Hayes, in West London are reputed to have heard a V1 cut out but no explosion. The authorities apparently checked out the area but couldn't find it.[3] And a similar account relates to the Edmonton area of East London. Two brothers were searching for shrapnel during their summer holidays in 1944, in the Town Road area, when they heard an engine sound they had not heard before. Looking up, they saw their first V1. As they watched, the engine cut out and a man in a local café shouted to them to come inside. The two brothers and the café owner dived under the table, but to their surprise the explosion never came. They assumed the V1 had landed, without going off, in a sewage farm some 300 yards away.[4]

A V1 is supposed to have also come down unexploded just south of the Thames in late summer 1944, after it hit a barrage balloon cable over what is now the A226 Thames Way road, near Swanscombe, Kent. It was thought to have been deflected into waterlogged ground near the Thames at Northfleet.[5]

Over the years the River Thames has proved to be home to a number of unexploded bombs from the Second World War and one might imagine it quite possible for a V-weapon to disappear beneath the water without going off. However, the only report found of an unexploded V1 in the Thames was from Able Seaman Douglas Rubery. He spent some time in London while his ship, the *Norse Lady*, was in dock at Poplar for repairs, having been damaged in an Atlantic convoy collision. One day he was on watch when he saw a V1 coming down towards the ship. He recalled that it hit the water behind the ship on its port side without exploding.[6]

Around the Thames estuary there were a couple of reports of flying bombs. John Sewell served as an armament officer at the Eastchurch airfield on the Isle of Sheppey, on the south side of the estuary. He recalled that one Sunday while his wife was preparing lunch they heard a V1. The engine stopped but no explosion followed. That evening John went to the local pub at Leysdown and found a bomb disposal squad having a drink there. Chatting to them, he found out that an old man had been walking his dog earlier and had gone into some bushes for a call of nature, only to come across an unexploded flying bomb. The squad was there to deal with it.[7] This account appears to be supported by another, regarding a conscientious objector by the name of Tom Chantrell. He served with a Royal Engineers BD squad in Tunbridge Wells and is said to have worked on a V1 that had crashed just outside Leysdown, on Sheppey. The V1 had apparently gone into the soft marshy ground at an awkward angle. It took several days to dig it out and remove the fuzes. The squad's officer was not in attendance at the site as he was on leave, his wife

Conscientious objector Tom Chantrell (second from left), was a member of the Non Combatant Corps and had volunteered for bomb disposal. He was attached to a Royal Engineers Bomb Disposal Section and helped recover an unexploded V1 from the Isle of Sheppey, Kent. He later made a name for himself as an artist in the film industry producing posters for films that included 'Jules Verne's Rocket to the Moon' and 'The Enemy Below'. (*http://www.chantrellposter.com/biography*)

being heavily pregnant at the time, but he was later presented with an OBE for successful efforts in relation to the incident. From then on Tom Chantrell would describe the OBE as Other Buggers' Efforts.[8]

On the Essex side of the estuary it is claimed that another V1 came down – and may still be there. On the parish history webpage there is a paragraph relating to an incident in WW2. It says:

> A German V1 pilotless flying bomb 'doodlebug' crashed into the sea wall this side of the Brandy Hole saltings near Beckney Farm. The V1 had either run out of fuel, or it had been shot down, or nudged down by an RAF fighter. The wrecked V1 must have skimmed the ground before colliding with the sea wall. Fortunately it did not explode. It was left embedded in the sea wall because it was an unexploded bomb and it was unwise to risk explosion or weakening the sea wall by digging it out. So, the aircraft is believed to be still there in the earthworks. This V1 was seen coming down with its engine off by Lawrence Copeland who lives in Golden Cross Road. He was in the small field at the bottom of Ashingdon Hill on the right on the corner of Canewdon Road. He and his friends were playing cricket on a Saturday and they all dived to the ground in case the flying bomb exploded.[9]

In a couple of books, *Britain under Fire* and *Lincolnshire Air War*, there is mention of a V1 that crashed into the mud of the Humber estuary without exploding. On Christmas Eve 1944, a member of the Civil Defence, Mr K. May of Winterton, saw the flame from the engine as the V1 flew up the Humber River near Ferry Sluice.[10] It was estimated that 40-plus V1s were launched that night from aircraft over the North Sea, targeting Manchester. Fishermen reported that quite a few of the V1 engines failed to start and they fell into the sea. The missile that Mr Kay saw, however, went into a spiralling flight-path, losing height until it eventually hit the mud of the estuary quite close to the eastern edge of Reads Island. The engine was apparently still running even after it came to rest sitting in the mud.[11] A V1 hitting the ground with its engines still running was not unknown. On 25 June 1944 one hit Kingsland Farm wood, in Kent, its engine going right up to the point of detonation. Later the engine of this V1 was found to contain bullet holes.[12]

It is possible some of the flying bombs from the Christmas Eve Manchester attack remain to be discovered at some point in the future – they could still be lying beneath the estuary's mudflats or under the North Sea.

Some celebrities have talked of the V1s they saw back in the war, a couple mentioning unexploded examples. In a 1997 interview with TV personality Bob Holness (the presenter of quiz shows 'Call my bluff' and 'Blockbusters'), he told of his childhood growing up in Ashford, Kent, in the heart of what was known as 'Bomb Alley'. He took an avid interest in aircraft and spent his spare time cycling around Kent visiting the sites of crashed aircraft and unexploded bombs. He was a member of the National Association of Spotters' Clubs at school. He recalled how he and a friend Don Nowers cycled off to look at an unexploded V1. At the crash site, they found the authorities were not present and the bomb was sitting unattended in the middle of a field. He went on, 'I unscrewed the detonators from the back and put them in my saddle bag.' His friend then suggested that it probably wasn't a great idea to collect such things, so they threw them into a field.[13]

Holness was not the only television personality to have taken an interest in V1s. The TV botanist, David Bellamy, also managed to obtain a rather dangerous component from one of Hitler's Vengeance weapons. Bellamy lived in St Dunstan's Hill, Cheam, during the war, and like most boys of the era enjoyed collecting shrapnel and other souvenirs. He wrote in his autobiography, 'We always hoped we would find the tail fin or something else recognisable from a doodlebug, so when we found a whole one that hadn't exploded on impact we were overjoyed. Although it hadn't gone off, its chambers had burst open so we carefully removed some of the explosive, filling our supply of two-pound jam jars ready to make fireworks.'[14] It is possible that David Bellamy was referring to the V1 mentioned in Chapter 2 that came down on 16 August 1944, near Addington Village Road, Croydon. This V1's warhead broke open on impact and the location is the nearest known UX V1 incident to Bellamy's home, though it is a good bike ride away.

Bellamy had already used his chemistry skills to make fireworks and had produced magnesium powder by filing down incendiary bomb casings. The V1 explosive was considered a bonus! Unfortunately while the fireworks were under construction in a friend's bedroom, there was an accident. An explosion blew out all the windows and resulted in the police, fire brigade and concerned neighbours all showing up at the house. Luckily, there were no lasting injuries.[15]

It is interesting to note that to assist with his firework-making, David Bellamy had borrowed a book from the library, *Brock's Own Book of Pyrotechnics*. The Brock family actually had some involvement with bomb disposal in the war, owing to their expertise with things that go bang.[16] For example, the Earl of Suffolk's bomb disposal experimental unit conducted tests in October 1940 with 'Mr Brock' at the Brock's Fireworks premises at Hemel Hempstead.[17]

The test involved filling a flower pot with thermite that was then placed on a bomb and ignited. The theory was that the thermite would burn through the hole in the bottom of the flower pot and then through the casing and ultimately begin burning the explosive materials. It was hoped that enough of the explosive in the bomb would be burnt to ensure a smaller explosion at the point it detonated, the best scenario being that all of it would burn and no explosion would occur. This process was expanded on as the war progressed.

Obtaining some sort of souvenir from an unexploded V1 was not an uncommon goal of schoolboys (and some adults) if they knew that one had come down locally. A 13-year-old schoolboy by the name of John Martin found that a V1 had come to rest unexploded on a recreation ground in his home town of Sevenoaks, Kent (thought to be the playing fields beside Hollybush Close and Hollybush Lane). The missile apparently sat close to the spot where a Ju-88 bomber had crash-landed earlier in the war. A BD squad, having made the warhead safe, took it away, but left the remainder of the wreckage in situ for some time. John Martin described the V1 as being broken open, which gave access to two compressed-air bottles that were wrapped in wire. Pieces of this wire were cut off by John and his brother with the intention of using

Looking east at the Sevenoaks recreation ground where an unexploded V1 is said to have come to rest. Behind the hedge on the right is the bowls club. Their website makes mention of the bomber that crashed close to the bowling green but unfortunately not the V1. (*Author collection*)

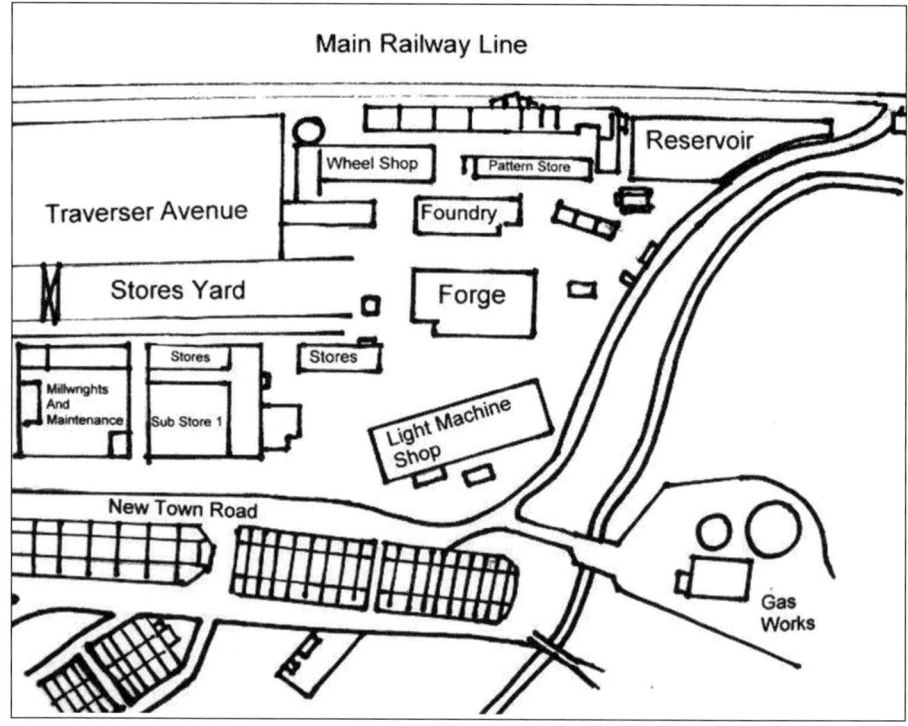

Plan of Ashford rail works as it was in 1942. An unexploded V1 was said to have come down near the reservoir shown top right. The reservoir has now gone and a cycle path crosses the area. It joins New Town Road with Hunter Avenue on the other side of the main railway line, through a tunnel at top right of the plan. The rail works had experience of a UXB prior to the V1, when on 30 September 1942 a Ju-88 attacked at low level, dropping four 250kg bombs, one of which failed to explode. (*Author collection*)

it as sprung undercarriage on the model rubber band-powered aircraft they were in the habit of constructing back then.[18]

Souvenirs taken from another UX V1 included a 'fuze plug' and other fuze related parts. These were liberated by Charlie Payne, an employee of Ashford Rail Works, where the V1 had apparently come down. The exact spot is supposed to have been near the reservoir at the end of the works site. Charlie was a member of the works' Home Guard unit and was tasked with guarding the missile until it had been made safe by a bomb disposal squad. It was after they had completed their work that he managed to obtain a few souvenirs, including said parts. He ended up giving them to a friend.[19]

Another unexploded V1 was said to have been seen near Wrotham, Kent. There was a barrage balloon site at the top of the hill where Wrotham Transmitting Station now stands and a schoolboy who was staying locally was a regular visitor. The personnel there one day took him to see a V1 that was apparently in quite fair condition, having come down without exploding a short distance away, near Old Coach Road. The M20 junction 2 exit sliproad is where the V1 was supposed to have actually hit the trees, where the ground rises to form part of the North Downs. The boy was later told that the V1 was dismantled and taken away to be displayed to raise money 'for Spitfires'.[20]

A former resident of Ealing, Roy Bartlett, recalled an unexploded V1 that fell in West London that he claimed was 'hushed up'. He worked at the huge AEC factory that was located close to the junction of Windmill Lane and the Uxbridge Road. While cycling to work early one morning, he was confronted by a barrier across the Uxbridge Road where it passed the Hanwell Mental Asylum. A warden told him there was an unexploded flying bomb in the grounds of the asylum but that as he worked at the factory further up the road he could go through. As he cycled past the rubble from the wall of the asylum that had been knocked down he saw a lorry in the grounds with a crane, lifting the battered but relatively intact V1.[21]

There is some evidence that a flying bomb did in fact hit the laundry of the asylum at Hanwell one Saturday afternoon, resulting in a number of casualties. But did it actually explode? Did Roy Bartlett really see the remains of an unexploded flying bomb or the substantial wreckage from an exploded one? The wall he saw smashed was to the north of the hospital buildings and it is unlikely that the V1 would have come from that direction, so it would be an easy assumption to make that it had been knocked down by the blast of an explosion. Or could there have been two separate incidents at the hospital? He claimed that work to repair the broken wall started the same day and that very few people at the factory were even aware of the incident.

It is quite easy to see how people could be misled into believing that they knew the locations of unexploded V-weapons. A report written in the war of an incident at Three Oaks, in Sussex, where a V1 exploded, having been shot down by an aircraft on 4 July 1944, clearly states 'The fuselage, in reasonably good condition has been found at Little Maxfield, Guestling – UXB notices posted.'[22] There was no UXB as such, but if you saw the signs and you saw the wreckage being recovered you could be forgiven for assuming it was a missile that hadn't gone off. This particular flying bomb certainly did go off, as the same report mentions three cottages and four houses were slightly damaged by the blast and a crater 8 x 24ft was created.[23]

It is not just V1s that people claim to have seen unexploded. An account was published in the book *Britain under Fire* in which Londoner Donald Ketley, a 13-year-old schoolboy at the time, remembered going past a house one morning where a V2 had just landed without exploding. He described how 'the monster had buried itself upright in someone's garden: the man of the house was standing, open-mouthed, in his doorway, staring in amazement at this enormous gleaming tower which had suddenly appeared in his yard.' Could he be possibly referring to the one that came down at 45 Northumberland Avenue, Hornchurch?

As the wartime generation passes away, so those stories of suspected unexploded V-weapons are lost. On the Continent too people still remember tales relating to V-weapons that may be awaiting rediscovery. One man, Anton Tibben, then aged twelve, was witness to a possible unexploded V1. At that time he was living at Breebrocksweg, between Heeten and Lettele in the Netherlands. He recalled a flying bomb that crashed to earth in countryside behind his house. A hole 5ft deep was found. Anton helped his father fill it in, as his father believed the missile had buried itself deep in the ground and being away from any buildings it would be safe to be left there.[24]

In 2005, Anton was reminiscing with his brother about the war and they came to the conclusion they should probably now tell the authorities about the possible V1. After all, though the land was still in the family, it might not always be and this particular V1 tale might be lost to history. The authorities scanned the area to a depth of 4.5m but the results were not entirely conclusive. It appeared there was more copper and zinc content in the ground than was usual but no traces of kerosene, the V1's fuel.[25] The local authorities would have had to find the money to pay for an excavation and it is not known if they ever did, or whether this V1 will remain one of those that lives on in local folklore.

One V1 that has entered into folklore, and indeed become an urban legend, is the 'Mystery of the Iron Lady'. In March 1945 a V1 is thought to have hit IJzeren Vrouw (Iron Lady) lake, at Hertogenbosch, in the Netherlands. Local opinion is divided as to whether or not it actually exploded. Some believe the lake was in fact hit by two V1s – Brothers Herman and Piet van Vugt lived 50m from the lake at that time and recalled that two bombs fell within a month of each other, one of which failed to go off. Sport divers have explored the lake on a number of occasions and it has been said that the tail end of a V1 was found during the 1960s, but this could easily have been from an exploded example.[26] It is a bit like the Loch Ness Monster – a number of believers but little evidence of this particular 'sleeping' monster.

Epilogue

This book has focused on the V-weapons' technical failings. However, these weapons can in fact be considered a failure on all levels. The whole point of their development was to turn the tide of war in the Nazis' favour: this obviously didn't happen. History has proved that the only things V-weapons did was to cause terror and grief. Many people were killed as a result of their production, development and use and much was destroyed. A typical example is the first one to land in the UK which caused the death of three people, including Rosemary Clarke, a three-year-old killed in her cot.[1]

The V-weapons succeeded only in escalating warfare to a whole new and frightening reality, where a warhead could be delivered across continents at the press of a button. Some may argue that the V1 and V2 played a crucial part in the evolution of weapons so frightening that they became a deterrent to their own use – a positive thing that has meant no 'world wars' since their creation. But are the modern day relatives of these missiles really a deterrent against war – surely there are extremists in the world who, given the chance, would use such weapons? Perhaps, unlike Hitler, they have yet to acquire the power to access such an arsenal.

So was there anything positive to come from the Nazi rocket programme? One cannot deny that the V-weapons were a fantastic leap in technology – but at such a price. Would man have reached the moon anyway and attained all its associated benefits, such as satellite technology, without the aid of the Nazi scientists and their 'progress at any cost' mentality? Yes, of course – it probably would have taken a little longer but the human race would surely have reached the same technology in time, hopefully with morally superior aspirations to those of the Nazi scientists. At the end of the day, despite what they may have claimed after the war, the German scientists involved with the *Vergeltungswaffen* (reprisal weapons) programme were well aware that their 'project' was supported by slave labour working in appallingly inhumane conditions. They also knew their work was being financed by the Nazi regime and its sole purpose was to deliver a warhead indiscriminately to the general populace of those who opposed them.

202 Disarming Hitler's V-Weapons

The conduct of the victors can also be questioned in that they willingly gave a relatively comfortable lifestyle to some of these scientists after the war, in exchange for their knowledge. A comfortable lifestyle was certainly not enjoyed by the V-weapon forced labour at places such as Mittelbau-Dora. It has been estimated that 12,000 forced labourers were killed during the production of V2s – more than were killed by its use.[2]

Despite the 'free labour', making V-weapons, particularly the V2, was a huge drain on resources and was the single most expensive development programme the Nazis embarked on. Each rocket cost as much to produce as a fighter aircraft. In some ways this helped the Allied war effort – Germany needed more aircraft yet the resources were being diverted to rockets. Even the fuel and explosives used for the rockets had a negative knock-on effect to other areas of the German war effort. Apparently to distil the fuel alcohol required for one V2 launch, some 30 tons of potatoes were needed, at a time when German civilians were going hungry.[3]

The Allies were well aware of the logistical problems the V-weapons programme presented to the Germans. In fact the Americans had thoroughly explored the possibility of copying the V1s and, rather ironically, using them

Bodies discovered at Mittelbau-Dora. This image shows the fate of some slave labour involved in the V-weapons programme. (*United States Holocaust Memorial Museum*)

to finish off the country that invented them. The parts from crashed V1s along with other intelligence gathered were quickly flown to Wright Field, Dayton, Ohio, to the headquarters of the Engineering Division of Air Technical Service Command. Here a team of expert technicians was ready to exploit their finds to the full.

From the components recovered the Americans were able to reverse-engineer the V1 and create their own version called the JB-2 (Jet Bomb 2). It was said that within 17 days of the wreckage arriving in the United States a working copy of the Argus engine was undergoing wind tunnel tests.[4]

By the end of December 1944, 2,000 of these JB-2s had been contracted to be built by the Republic Aircraft Corporation at their Long Island plant. The work of building the engines was farmed out to the Ford Motor Company and Jack Heintz Incorporated was to build the auto-pilots.[5] Experimental launches took place at Muroc Dry Lake in the Mojave Desert, which in years to come would become a landing site for the Space Shuttle. Following these tests, ramps similar to the German ones constructed in Europe were utilized at a test site at Elgin Field, Florida.[6] The ramps had originally been built for training purposes, to look at the best ways to attack and destroy them. Test firings from ramps were also conducted at Holloman Air Force Base, Alamogordo, New Mexico. In addition, JB-2s were air-launched from B-17, B-24 and B-29 bombers.

The technical challenges were not really a problem for the Americans as regards using the JB-2 against Germany. However, as the Germans had already discovered, the logistics of such attacks were cause for concern. Calculations were made and the provision of operational and service personnel was seen to be a major obstacle. It was estimated that for worthwhile results the Allies would have to launch 100 bombs a day, which would require over 5,500 personnel.[7] These men would have to come from resources back in the United States, so as not to deplete forces already on the ground in Europe. Other considerations included the fact that about 70 ramps would need to be built and the worry, which mirrored the Germans' concerns, that explosives needed for the JB-2s would impact on the supply of artillery shells and aerial bombs, at a time when they couldn't be spared. Space in cargo ships across the Atlantic would also be put under pressure. With all these problems and with the war moving quickly on, squeezing the Germans back into their homeland, the JB-2 launches never actually became a reality. It is interesting to note that possible objectives in Japan were also considered as targets for JB-2s, though where they would have been launched from is open to speculation.[8]

The German V-weapons were simply vehicles for transporting a warhead, in the same way a bomber was, but a bomber could hopefully be used more than once. As a means of delivery, the V2 had in its favour the speed at which it travelled and its mobility of launch sites, making counter measures virtually impossible. Although some 7,250 people were killed by the 3,000-plus V2s launched against the Allies, if the cost and effort had been spent on producing more conventional weapons instead the toll might have been even greater. The same could be said for the V1s, since they too were not highly effective. It has been estimated that only 25 per cent of V1s launched hit their target.

V1s didn't have the speed or mobility of launch sites that the V2s had, and it was therefore much easier to develop counter measures against them. Air-launching from bombers took care of the mobility problem but they still had a failure rate of around 40 per cent. An added problem was the fact that on launch the weapon lit up the bomber that had been carrying it. Allied night fighters could then take advantage of their improved view. It has been estimated that 65 to 77 bombers failed to return from V1 launch missions for one reason or another.[9]

So if the resources had been put into more conventional weapons it may be argued that the Germans could have disrupted the Allied war effort to a much larger degree. The British Government, not to mention the bomb disposal community, was somewhat surprised for instance, that the Germans did not make more use of butterfly bombs. Cheap to produce, these small anti-personnel 'cluster bombs' could be dropped in their hundreds by one aircraft. One particular raid on the city of Hull using butterfly bombs had great effect. Though they obviously did not have the destructive power of a one-ton warhead, they still had the potential to effectively paralyse communications and industry. The fuzes these bombs employed included ones that became sensitive to vibration after having come to rest, while others were timed to detonate after a fixed period. Considered too dangerous to disarm, the British policy was to dispose of them by blowing them up in situ, after first taking precautions to limit the damage by surrounding them with sandbags for example – not so easy when they were found hanging from gutters or telephone lines. Dealing with butterfly bombs was an extremely time consuming job and delayed the populace in getting back to a normal routine of work. Their effectiveness was of such concern that the British made sure that news reports of them were heavily censored. Had they been dropped in large numbers on major cities, or the locations where there was a build-up of troops in preparation for D-Day, then the results could have been devastating.

The Germans could also have put more resources into developing a good long-range heavy bomber. Had they done so they would have given themselves

Butterfly bombs in the author's collection. These cluster bombs were an effective weapon in Hitler's arsenal and the British authorities feared that they would be used in much larger numbers than they actually were. (*Author collection*)

more options for delivering accurately either the one-ton bomb, as used in a V-weapon, or hundreds of smaller bombs.

In summary, Hitler's original 'wonder weapons' did not alter the course of the war. Ultimately they became nothing more than defunct trophies for the victorious Allies to be photographed next to. They are now relegated to harmless museum pieces, used to educate today's generation about the lengths an evil idealism will go to in the quest for power.

And what became of the main personalities involved in this story? On the German side it is acknowledged that, next to Hitler himself, Wernher von Braun was the most prominent driving force behind the rocket building programme. In 1945 the former Nazi, afraid of being captured by the Russians, gave himself up to the Americans. Von Braun went to Fort Bliss near El Paso in the United States, where he worked as part of a team refurbishing captured V2s for launching at the White Sands Proving Ground in New Mexico.

Antwerp, 1945 (*Author collection*)

The American government glossed over Von Braun's Nazi past and in 1950 he was transferred to Huntsville, Alabama, where he became involved in the 'space race'.[10] He was given US citizenship in 1955. Von Braun's team went on to play a part in the launch of the West's first satellite in 1958 and later he became involved in the Saturn Rocket project that culminated with man landing on the moon. Von Braun retired from NASA in 1972 and died a natural death in 1977, aged 65.

In Britain, John Pilkington Hudson was without doubt the key player in the Second World War when it came to disarming V1s. He truly led from the front, devising tools and methods for dealing with all manner of unexploded German weapons. Regularly he would put his own life at serious risk while perfecting his ideas. This development work did not always take place in comfortable laboratory conditions, but would often occur at the bottom of muddy holes with live weapons containing completely unknown but very real threats. His discoveries and ideas saved not only his own life but also the lives of many other BD personnel. It has been said that he personally made sure his disarming methods were sound at least twice before asking anyone else to try the same thing.

After the war John Hudson returned to his pre-war interest of horticulture and for a time worked in New Zealand's Department of Agriculture. He returned to the UK in 1948 and went on to become a professor with the University of Nottingham, studying the effect of the environment on plant growth. In 1967 he took up the Directorship of Long Ashton Research Station in Bristol University. He retired eight years later. In December 2007 he passed away peacefully at his home in Wrington village, near Bristol, at the age of 97.[11]

Named in his honour, 'Hudson House', has been the South London home of a Royal Engineers Territorial Army Bomb Disposal unit. Into the twenty-first century they, along with other BD forces, continue to apply similar ingenuity, physical stamina and courage while dealing with the threat from modern day improvised explosive devices.

Notes and Sources

Introduction
1. *Steady as you go: A Canadian at sea*, Hugh Cronyn GM, privately printed, p.14.
2. *Unexploded Bomb*, Major A.B. Hartley MBE, RE, Cassel & Co Ltd, London, 1958, illustration opposite p.72.
3. *Conversations with Marco Polo*, Mark Denny and Joanna Lee Nelson, Xlibris Corporation, USA, p.107.
4. *The V2 and the Russian and American Rocket Program*, Claus Reuter, S.R. Research and Publishing, 2000, p.43.
5. *Hitler's Rocket Soldiers*, Murray R. Barker and Michael Keuer, Tattered Flag Press, Pulborough, 2011, pp.102,122,149–50.
6. http://www.ushmm.org/wlc/en/article.php?ModuleId=10005322 and http://www.v2rocket.com/start/chapters/mittel.html.
7. *Odyssey of a Jewish Sailor*, Captain F. Ashe Lincoln QC, RNVR, Minerva Press, London, p.20.
8. AVIA 22/2463 *German flying bombs: reports*, The National Archives, UK.
9. AVIA 15/3172 *Flying bomb defence: idea submitted by Roy Law, a nine year old boy*, The National Archives, UK.
10. HO 186/2819 *BOMBS, Disposal of unexploded flying bombs*, The National Archives, UK.
11. AVIA 11/49 *Proposal to block flying bomb vents with paper*, The National Archives, UK.
12. *Service Most Silent*, John Frayn Turner, George G. Harrap and Co Ltd, London, 1955, pp.79–85 and *Secret Naval Investigator*, Commander Ashe Lincoln QC, William Kimber, London, 1961, pp.98–9.
13. *Service Most Silent*, Ibid, pp.85–93 and *Secret Naval Investigator*, Ibid, pp.98–9.
14. *Dragons can be defeated*, Major D.V. Henderson, F.M. Spink and Son Ltd, 1984, pp.89, 92 and 29.
15. AIR 2/9224 *German long range rockets: disposal*, The National Archives, UK.
16. *Hitler's Rocket Soldiers*, Ibid, p.118.

Chapter 1: June 1944: The First V1s
1. DEFE 40/14 *Examination of German V2 rockets found in Sweden*, The National Archives, UK.
2. *Rocket*, Air Chief Marshal Sir Philip Joubert de la Ferte, Hutchinson, London, 1951, pp.50–4.
3. *Ibid*, pp.50–4.

4. DEFE 40/14, *Ibid.*
5. *Man who saved London: The story of Michel Hollard*, George Martelli, Companion, London, 1960.
6. *V-1 Flying Bomb 1942–52*, Steven J. Zaloga, Osprey, Oxford, 2005, p.18.
7. AVIA 22/2463 *German flying bombs: reports*, The National Archives, UK.
8. AIR 29/887, *Ibid. Operations Record Books, Miscellaneous Units*, The National Archives, UK.
9. AVIA 22/2463 *Ibid.*
10. HO 199/465 *Flying bombs: details of incidents technical reports and instructions on dealing with those that fall without exploding*, The National Archives, UK.
11. AIR 27/765 *No 96 Squadron: Operations Record Book*, The National Archives, UK.
12. Sussex Police Authority records, East Sussex Records Office.
13. AVIA 22/2463, *Ibid.*
14. *Ibid.*
15. AVIA 22/2455 *Removal of explosives from unexploded bombs by dissolving in solvent*, The National Archives, UK.
16. *The Danger of UXBs*, Lieutenant Colonel E. E. Wakeling, B.D. Publishing, Bourne End, Bucks, 1996, p.50.
17. WO 373/69 *Non-combatant gallantry awards*, The National Archives, UK.
18. *Unexploded Bomb*, Major A. B. Hartley MBE, RE, Cassel & Co. Ltd, London, 1958, pp.141–4.
19. *Dragons can be defeated*, Major D. V. Henderson, F.M. Spink and Son Ltd, 1984, p.52.
20. Correspondence from John Hudson via his son to author, January 2006.
21. Sussex Police Authority records, *Ibid.*
22. Correspondence from John Hudson, *Ibid.*
23. WO 195/743 *Research and Development Sub-Committee of Unexploded Bomb Committee: radiographic examination of German 500 Kilogram bomb with covered fuze*, The National Archives, UK.
24. WO 195/426 *Research Committee of Unexploded Bomb Committee: immunisation of German fuzes by X-Rays*, The National Archives, UK.
25. WO 195/509 *Research and Development Sub-Committee of Unexploded Bomb Committee: research on immunisation of R.M. fuzes by X-Rays at Royal Aircraft Establishment, Farnborough, Hants*, The National Archives, UK.
26. *Ibid.*
27. http://course.wilkes.edu/HWkoch/.
28. AVIA 22/2463, *Ibid.*
29. *Ibid.*
30. AVIA 15/249 *BOMB DISPOSAL: Methods of dealing with unexploded bombs by RAF.*
31. AVIA 22/2463, *Ibid.*
32. *Buzz Bomb Diary* D.G. Collyer, Kent Aviation Historical Research Society, Deal, Kent, 1994, p.72.

33. WO 195/1057 *Unexploded Bomb Committee: explosion during trials of acid cutting into bomb cases*, The National Archives, UK.
34. AVIA 22/2463, *Ibid*.
35. *The Danger of UXBs, Ibid*, p.55.
36. *Unexploded Bomb, Ibid*, p.46.
37. Correspondence from Sergeant Fred Harding's granddaughter to author, 2010.
38. Correspondence from John Hudson, *Ibid*.
39. *The Times*, obituary, 7 June 1996.
40. HO 199/367 *Unexploded flying bombs*, The National Archives, UK.
41. AVIA 22/2463, *Ibid*.
42. HO 199/367, *Ibid*.
43. *Ibid*.
44. Correspondence from Tom Ledger to author, 2011.
45. HO 198/84 *Bomb Census Papers FLYING BOMBS (V 1s), London*, and HO 199/367, *Ibid*.
46. AVIA 22/2463, *Ibid*.
47. WO 195/6812 *Unexploded Bomb Committee: disposal of unexploded flying bombs*, The National Archives, UK.

Chapter 2: More V1 'Duds'
1. *Buzz Bomb Diary,* D.G. Collyer, Kent Aviation Historical Research Society, Deal, Kent, 1994, p.149.
2. *The Diary of an ARP Warden*, E.J. Carter, Waltham Abbey Historical Society, p.24.
3. AIR 27/955 *No 137 Squadron: Operations Record Book: Appendices*, The National Archives, UK.
4. Sussex Police Authority records, East Sussex Records Office.
5. AIR 29/1008 *Operations Record Books, Miscellaneous Units No. 49, Faygate,* The National Archives, UK.
6. HO 198/86 *Bomb Census Papers FLYING BOMBS (V1s), London*, The National Archives, UK.
7. Correspondence from Doreen Ashmeade to author, April 2010.
8. Correspondence from Ray Scott to author, October 2010.
9. WO 195/6812 *Unexploded Bomb Committee: disposal of unexploded flying bombs*, and AVIA 22/2463 *German flying bombs: reports*, The National Archives, UK.
10. *The War in East Sussex*, Sussex Express and County Herald, Lewes, 1985.
11. WO 195/6812, *Ibid* and AVIA 22/2463, *Ibid*.
12. HO 199/367 *Unexploded flying bombs*, and HO 186/2819: *BOMBS, Disposal of unexploded flying bombs*, The National Archives, UK.
13. http://www.memorywall.org.uk/page_id__504_path__0p1p65p232p233p236p84p210p.aspx, David Taft.
14. WO 195/6812, *Ibid*.
15. HS 7/16 *"V" Weapon section: CROSSBOW and BIG BEN; operations in Normandy; 6 Airborne division; VARSITY operation*, The National Archives, UK.

Notes and Sources 211

16. *Ibid.*
17. HO 199/465 *Flying bombs: details of incidents technical reports and instructions on dealing with those that fall without exploding*, The National Archives, UK.
18. AIR 27/8 *No 1 Squadron: Operations Record Book: Appendices*, The National Archives, UK.
19. HO 199/367 *Unexploded flying bombs*, and HO 186/2819 *BOMBS, Disposal of unexploded flying bombs*, The National Archives, UK and *Buzz Bomb Diary, Ibid*, p.152.
20. AVIA 22/2463, *Ibid.*
21. WO 195/6812, *Ibid.*
22. AVIA 22/2463, *Ibid.*
23. *Buzz Bomb Diary*, D.G. Collyer, Kent Aviation Historical Research Society, Deal, Kent, 1994, p.152.
24. *Diver! Diver! Diver!*, Brian Cull with Bruce Lander, Grub Street, London, 2008, pp.238–9.
25. *Ibid.*
26. BBC People's War Article ID A4388826.
27. *Buzz Bomb Diary, Ibid*, p.27.
28. BBC People's War, Article ID A6871034.
29. *Ibid*, Article ID A5106223.
30. Correspondence from Bert Clinch to author March 2011.
31. *Ibid.*
32. *Designed to Kill*, Major Arthur Hogben, Patrick Stephens, Wellingborough, Northamptonshire, 1987, pp.164-5.
33. www.thecivicsociety.org/heritage_days.
34. Correspondence from Bert Clinch to author, *Ibid.*
35. *Designed to Kill, Ibid*, pp.164-5.
36. *Unexploded Bomb*, Major A. B. Hartley MBE, RE, Cassel & Co. Ltd, London, 1958, p.188.
37. AVIA 22/2463, *Ibid.*
38. Sussex Police Authority records, *Ibid.*
39. *Ibid.*
40. HO 199/367 *Unexploded flying bombs*, The National Archives, UK.
41. AIR 27/955 *No 137 Squadron: Operations Record Book: Appendices*, The National Archives, UK.
42. *Buzz Bomb Diary, Ibid*, p.53.
43. AIR 27/935 *No 129 Squadron: Appendices*, The National Archives, UK.
44. AVIA 22/2463, *Ibid.*
45. HO 199/367, *Ibid.*
46. *Ibid.*
47. *Ibid.*
48. *Ibid.*
49. *Ibid.*
50. *Air Launched Doodlebugs*, Peter J.C. Smith, Pen & Sword Aviation, Barnsley, 2006, p.195.

51. HO 199/367, *Ibid*.
52. *Air Launched Doodlebugs, Ibid*, p.93.
53. *The Danger of UXBs*, Lieutenant Colonel E.E. Wakeling, B.D. Publishing, Bourne End, Bucks, 1996, p.83.
54. *The Danger of UXBs, Ibid*, pp.91-2.
55. Correspondence from Wally Fielding (ex 220 Sec 22 BD Coy RE) to author, 1990.
56. *Highly Explosive*, John Frayn Turner, George G. Harrap and Co Ltd, London, 1961.
57. *Air Launched Doodlebugs, Ibid*, p.92.
58. HO 198/77 *Bomb Census Papers FLYING BOMBS (V1s), Cambridge and WO 166/14797 War Diaries, 138 H.A.A. Regt*, The National Archives, UK.
59. *Air Launched Doodlebugs, Ibid*, p.124–5.
60. *Ibid*, p.202.
61. HO 198/77, *Ibid*.
62. *Air Launched Doodlebugs, Ibid*, p.147.
63. *Ibid*, p.177.
64. HO 199/367, *Ibid*.
65. *Bombs and Boobytraps*, Captain H. J. Hunt, Romsey Medal Centre, 1986, p.33.
66. HO 198/78 *Bomb Census Papers FLYING BOMBS (V1s), Cambridge*, The National Archives, UK.
67. *Ibid*.
68. Private papers of Brigadier G. Streeten CBE, MC, Imperial War Museum document 7475.
69. HO 199/465 *Ibid*.
70. *Unexploded Bomb, Ibid*, p.184.

Chapter 3: Unexploded V2s in England

1. AVIA 6/25656 *Report on V2 which fell in Sweden on 13 July 1944 and other papers*, The National Archives, UK.
2. http:www.df.lth.se/~triad/rockets/therocket.html and AVIA 6/25656, *Ibid*.
3. DEFE 40/18 *German rocket development: report on test firings in Poland*, The National Archives, UK.
4. *The Diary of an ARP Warden*, E. J. Carter, Waltham Abbey Historical Society, p.26.
5. Sussex Police Authority records, East Sussex Records Office.
6. AIR 2/9224 *German long range rockets: disposal*, The National Archives, UK.
7. *Ibid*.
8. AIR 2/9224, *Ibid*.
9. *Ibid*.
10. *The Diary of an ARP Warden, Ibid*, p.31.
11. HO 199/465 *Flying bombs: details of incidents technical reports and instructions on dealing with those that fall without exploding*, The National Archives, UK.
12. http://v2rocket.com/start/deployment/timeline.html.
13. *Unexploded Bomb*, Major A. B. Hartley MBE RE, Cassel & Co. Ltd, London, 1958, pp.191–2.

Notes and Sources   213

14. HO 198/104 *Bomb Census Papers, LONG RANGE ROCKETS (V2s), Cambridge*, The National Archives, UK.
15. AIR 20/7699 *V1s and V2s: reports and technical details*, The National Archives, UK.
16. *Unexploded Bomb*, Major *Ibid*, pp.192-4.
17. *Ibid*, p.193.
18. *Designed to Kill*, Major Arthur Hogben, Patrick Stephens, Wellingborough, Northamptonshire, 1987, pp.166–8.
19. AIR 2/9224, *Ibid*.
20. HO 198/104, *Ibid*.
21. *Ibid*.
22. *Unexploded Bomb*, *Ibid*, p.194.
23. *Ibid*, p.195.
24. HO 186/2381 *AIR RAIDS, Incident Reports, Deptford*, The National Archives, UK.
25. *Ibid*.
26. http://londonist.com/2009/01/london_v2_rocket_sitesmapped.phpandhttp://flyingbombsandrockets.com/V2_intro.html.
27. HO 186/2416 *AIR RAIDS, Incident Reports, Southgate*, The National Archives, UK.
28. http://forum.keypublishing.com/showthread.php?t=76992 and *The Flying Bomb War*, Peter Haining, Robson Books, 2002, centre pages.
29. http://www.v2rocket.com/start/deployment/timeline.html.
30. Correspondence from John Pridige, January 2006.
31. http://wwii-letters-to-wilma.blogspot.co.uk/2012/03/27-march-1945.html.

Chapter 4: Europe
1. http://www.cwgc.org Service No. 1285110, Eldred Francis MIDDLETON.
2. WO 171/1979 *War Diaries 226 Sec*, The National Archives, UK.
3. *Ibid*.
4. AIR 29/887 *Operations Record Books, Miscellaneous Units*, The National Archives, UK.
5. WO 171/1958 *War Diaries 24 Coy*, The National Archives, UK.
6. WO 171/1977 *War Diaries 224 Sec*, The National Archives, UK.
7. WO 171/1957 *War Diaries 23 Coy*, The National Archives, UK.
8. WO 171/1966 *War Diaries 53 Sec*, The National Archives, UK.
9. WO 171/1959 *War Diaries 25 Coy*, The National Archives, UK.
10. 951st Battalion Newsletter, Fire Mission, September 1998.
11. *Daily Mail*, 1945.
12. *Diver! Diver! Diver!*, Brian Cull with Bruce Lander, Grub Street, London, 2008, pp.375–6.
13. http://www.volksfreund.de/2221951.
14. *Ibid*.
15. *Ibid*.
16. *Ibid*.
17. http://users.skynet.be/fb061479/Ongevallen.htm.

18. http://www.hbavo-heemstede.nl/main.php?page=17.
19. HO 199/465 *Flying bombs: details of incidents technical reports and instructions on dealing with those that fall without exploding,* The National Archives, UK.
20. http://www.v2rocket.com/start/deployment/westerwald.html.
21. *Hitler's Rocket Soldiers*, Murray R. Barker and Michael Keuer, Tattered Flag Press, Pulborough, 2011, pp.206–7.
22. *Ibid*, p.148.
23. http://www.ww2museums.com/article/11956/Former-V2-Launch-Site-Rijsterbos.htm.
24. WO 219/4937 *Operation Crossbow: reports on visits to launching sites,* The National Archives, UK.
25. *Ibid*.
26. *Ibid*.
27. *Hitler's Rocket Soldiers, Ibid,* pp191-2.
28. *All Theirs*, Noel Cashford MBE, RNVR, ALD Design & Print, Sheffield, 2004, p.115; and *Designed to Kill*, Major Arthur Hogben, Patrick Stephens, Wellingborough, 1987, p.267.
29. AVIA 22/2463 *German flying bombs: reports,* The National Archives, UK.
30. *Captains of Bomb Disposal 1942–1946,* T. Dennis Reece, Xlibris Corporation, USA, p.62.
31. *Ibid.,* p.71.
32. AIR 2/9224, *German long range rockets: disposal* and AVIA 6/25652 *Big Ben Sub-Committee of Crossbow Committee: external reports,* The National Archives, UK.
33. AVIA 6/25652, *Ibid*.
34. AIR 29/886 Operations Record Books, Miscellaneous Units, The National Archives, UK.
35. WO 195/7601- *Unexploded Bomb Committee: recovery of unexploded warhead of German long-range rocket* and AIR 2/9224, *Ibid,* The National Archives, UK.
36. WO 195/7601, *Ibid*.
37. *Ibid*.
38. AIR 2/9224, *Ibid*.
39. WO 195/7674 *Unexploded Bomb Committee: warhead of German long-range rocket* and AVIA 6/25652, *Ibid,* The National Archives, UK.
40. *Ibid*.
41. WO 195/7673 *Unexploded Bomb Committee: bomb fuzing system of German long-range rocket,* The National Archives, UK.
42. WO 195/7602 *Unexploded Bomb Committee: recovery of unexploded warhead from flying bomb,* The National Archives, UK.
43. AIR 29/887, *Ibid*.
44. AIR 29/886, *Ibid*.
45. AIR 29/887, *Ibid*.
46. AIR 29/886, *Ibid*.
47. *Ibid*.
48. AIR 29/887, *Ibid*.
49. *V-1 Flying Bomb 1942–52*, Steven J. Zaloga, Osprey, Oxford, 2005, p.37, and http://www.V2Rocket.com.

Notes and Sources 215

50. AIR 29/886, *Ibid*.
51. *Ibid*.
52. *Ibid*.
53. AIR 29/887, *Ibid*.
54. AIR 29/886, *Ibid*.
55. AIR 29/887, *Ibid*.
56. *Ibid*.
57. http://www.V2Rocket.com.
58. *Ibid*.
59. AIR 29/887, *Ibid*.
60. *Ibid*.
61. AIR 29/59 *Operations Record Books, Miscellaneous Units*, The National Archives, UK.
62. *Ibid*.
63. *Ibid*.
64. *Ibid*.
65. AIR 29/887, *Ibid*.
66. AIR 29/886, *Ibid*.
67. AIR 29/887, *Ibid*.
68. AIR 29/886, *Ibid*.
69. *Ibid*.
70. *Ibid*.
71. AIR 29/60 *Operations Record Books, Miscellaneous Units*, The National Archives, UK.
72. Royal Engineers Association Bomb Disposal Branch Bulletins Spring 1995 pp.11–12, Christmas 1995 p.8 and No. 125, 2006 p.2.
73. *The Danger of UXBs*, Lieutenant Colonel E.E. Wakeling, B.D. Publishing, Bourne End, Bucks, 1996, pp.59,76,88.
74. *Ibid*, pp.95–6.
75. *Dragons can be defeated*, Major D.V. Henderson, F.M. Spink and Son Ltd, 1984, p.77.
76. *The Danger of UXBs, Ibid*, pp.191–2 and *Danger UXB*, M. J. Jappy, Channel 4 Books, London, 2001, pp.182–3.
77. AIR 29/59, *Ibid*.
78. AIR 29/887, *Ibid*.
79. AIR 29/59, *Ibid*.
80. AIR 29/887, *Ibid*.
81. AIR 29/59, *Ibid*.
82. *Ibid*.
83. ADM 116/4920 *Operation "Crossbow": evacuation of Admiralty staff, security, emergency plans, disposal of unexploded missiles, etc*, The National Archives, UK.
84. ADM1/30639 *Awards to 31 officers and men of Belgian and Dutch "P" Parties for services in mine clearance and disposal prior to and following close of war with Germany*, The National Archives, UK.
85. Conversation with Ray Maries, February 2010.

86. *Open the Ports*, J. Grosvenor & L.M. Bates, William Kimber & Co. Ltd, 1956, London, p.157.
87. Conversation with Ray Maries, *Ibid*.
88. ADM 1/30639, *Ibid*.
89. *Air Launched Doodlebugs*, Peter J.C. Smith, Pen & Sword Aviation, Barnsley, 2006, p.141.
90. 'A Mystery Solved', *Britain at War* Magazine, Issue 9, September 2009.

Chapter 5: Tools of the Trade
1. *A Photographic Story of Wartime Bomb Disposal*, Lieutenant Colonel E. E. Wakeling, B.D. Publishing, Bourne End, Bucks, 1995, p.14.
2. WO 195/1230 *Research and Development Sub-Committee of Unexploded Bomb Committee: effect of vibrations from mechanical digging tools on No. 17 clock*, The National Archives, UK.
3. *Bombs and Boobytraps*, Captain H. J. Hunt, Romsey Medal Centre, 1986, p.60.
4. *The Danger of UXBs*, Lieutenant Colonel E. E. Wakeling, B.D. Publishing, Bourne End, Bucks, 1996, p.90 and http://www.cwgc.org.
5. Correspondence from Wally Fielding (ex 220 Sec 22 BD Coy RE) to author, 1990.
6. *Ibid*.
7. *A Photographic Story of Wartime Bomb Disposal*, *Ibid*, p.22.
8. http://www.contaminatedland.co.uk/sere-dip/estd-uxb.htm.
9. *The Explosive Years*, Bert Blackmore Tom Donovan Publishing Ltd, London, 1994, pp.153-5.
10. AIR 2/9185 *Reports of operations on bomb disposal carried out by 0.10 (B.D.)*, The National Archives, UK.
11. WO 166/4000 *War Diaries, 8 Bomb Disposal Company*, The National Archives, UK.
12. *RAE Exhibition of Aircraft and Ground Electrical Equipment November 1945, Annotated Catalogue Supplement List of Secret Items on Disposal of Enemy Bombs*, memo EL.1343.
13. *Unexploded Bomb*, Major A. B. Hartley MBE, RE, Cassel & Co. Ltd, London, 1958, pp.54–6.
14. WO 195/299 *Unexploded Bomb Committee: fuze dischargers and keys*, The National Archives, UK.
15. *Ibid*.
16. AIR 2/9185, *Ibid*.
17. WO 195/4534 *Unexploded Bomb Committee: Zus.40 fuze immunisation by Stevens stopper and fuze extractor design III*, The National Archives, UK.
18. AIR 2/9188 *Technical reports on bomb disposal equipment*, The National Archives, UK.
19. WO 195/5009 *Unexploded Bomb Committee: progress report on search for universal bomb disposal liquid*, The National Archives, UK.
20. WO 195/4979 *Unexploded Bomb Committee: radiographic investigation of Stevens stopper technique used with Ardacol for fuze "Y"*, The National Archives, UK.

21. WO 195/4578 *Unexploded Bomb Committee: steaming-out bombs, temperatures reached in main filling and gaines during standard procedures*, The National Archives, UK.
22. *Life on the edge*, Peter Varey, PFV Publications, Cambridge, 2012, p.15 and *Unexploded Bomb, Ibid*, p.15.
23. AIR 2/9188, *Ibid.*
24. *Handbook of British Bomb Disposal Equipment for Dealing with German Bomb Fuzes*, BD Publishing, Bourne End, p.9.
25. *Unexploded Bomb, Ibid*, p.80.
26. http://www.drandrew.co.uk/lazy-tools-to-help-us/68/dowsing.
27. AIR 2/9188, *Ibid.*
28. WO 195/8436 *Unexploded Bomb Committee: pyrotechnic fuze extractors*, The National Archives, UK.
29. *Ibid.*
30. *Recollections of a Mustang*, Robert W. (Ike) Eigell, self-published, USA, 1994.
31. AIR 2/9188, *Ibid.*
32. AIR 2/9185, *Ibid.*
33. *History of US Navy Bomb Disposal*, USN EOD Association, 1992, Research programs p.1.
34. AIR 29/59 *Operations Record Books, Miscellaneous Units*, The National Archives.
35. *A Life in my Hands*, Wally Thomas, Heinemann, London, 1960, pp.1–3.
36. *Ibid.*
37. *Bombs and Boobytraps*, Captain H. J. Hunt, Romsey Medal Centre, 1986, p.83.
38. *Ibid*, p.74.
39. *Unexploded Bomb, Ibid,* p.110.
40. AIR 2/9185, *Ibid.*
41. ADM 1/11103 *Bomb and mine disposal duties: issue of badges*, The National Archives.
42. *Ibid.*
43. *The Story of the George Cross*, Sir John Smyth, Arthur Baker Ltd, London, 1968, p.23.
44. WO 195/7601 *Unexploded Bomb Committee: recovery of unexploded warhead of German long-range rocket*, The National Archives, UK.
45. *Service Most Silent*, John Frayn Turner, George G. Harrap and Co Ltd, London, 1955, p.40.
46. *Secret Naval Investigator*, Commander Ashe Lincoln QC, William Kimber, London, 1961, pp.85-9.

Chapter 6: Post-War Discoveries
1. *V1 Eifelschreck*, Wolfgang Gückelhorn & Detlev Paul, Helios, Germany, 2004, p.186.
2. *Ibid*, p.187.
3. *Ibid*, p.186.
4. AIR 29/887 *Operations Record Books, Miscellaneous Units*, The National Archives, UK.

5. AIR 29/887, *Ibid.*
6. AIR 29/886 *Operations Record Books, Miscellaneous Units*, The National Archives, UK.
7. AIR 29/887, *Ibid.*
8. AIR 29/886, *Ibid.*
9. *Wie ich die V1 entschärfte*, Spiegel 26/1951, Hamburg, p.14ff.
10. *Ibid.*
11. *Ibid.*
12. *Ibid.*
13. *V1 Eifelschreck, Ibid*, p.187.
14. http://www.piershil.com/index.php/oorlog/548-1945-1967v1s-te-piershil.
15. http://www.volksfreund.de/2221951.
16. *V1 Eifelschreck, Ibid*, p.189.
17. *Ibid.*
18. Newspaper *Cuxhavener Nachrichten*, 8 February 2003.
19. *V1 Eifelschreck, Ibid*, p.190.
20. http://www.defensie.nl/actueel/nieuws/2007/04/03/4686633/V1_resant_verwijderd_van_Schiermonnikoog
21. *V2 Gefrorene Blitze*, Wolfgang Gückelhorn & Detlev Paul, Helios, Germany, 2007, p.205.
22. http://www.volksfreund.de/2221827.
23. *For RAF Eyes Only*, Jim Jenkinson, Janus Publishing Co, London, 1993, p.83.
24. *Hitler's buried bombs threaten cleanup of Olympic's site*, http://www.bloomberg.com.
25. http://lse.co.uk, UK News, *Doodlebug Alert Proves To Be 'Concrete'*, 29 July 2007.
26. http://www.metro.co.uk/news/837951-ufo-spotted-on-the-bed-of-the-thames.
27. http://www.naylandconservation.org.uk/Achievements.html.
28. http://www.indianamilitary.org/FreemanAAF/Museum/FF_museum.html.
29. http://www.ostsee-zeitung.de/ozdigital/archiv.phtml?SID=75f2a66b3e2ab67f0b76cd3feeb058f3&param=news&id=3132400.
30. http://www.v2rocket.com/start/others/news.html.
31. http://www.rspb.org.uk/community/placestovisit/wallaseaisland/b/wallasea-island-blog/archive/2012/03/29/wallasea-project-gets-a-rocket.aspx.

Chapter 7: Fact or Fiction?
1. BBC People's War Article ID A6787308.
2. *Ibid*, ID A1980858.
3. Conversation with Gareth Owen, 2011.
4. BBC People's War Article ID A4018349.
5. *Ibid*, ID A2355978.
6. *Ibid*, ID A6903993.
7. Interview Audio Recording 12606, Imperial War Museum, London.
8. http://www.essexinfo.net/ashingdonparish/history/), http://www.chantrellposter.com/biography.

9. http://www.essexinfo.net/ashingdonparish/history/.
10. *Air Launched Doodlebugs*, Peter J.C. Smith, Pen & Sword Aviation, Barnsley, 2006, p.154.
11. *Ibid*, p.155.
12. Sussex Police Authority records, East Sussex Records Office.
13. *Focus*, Ministry of Defence magazine, September 1997.
14. *David Bellamy Jolly Green Giant*, David Bellamy, Century, London, 2002, p.73.
15. *Ibid*, p.83.
16. *Softly tread the brave*, Ivan Southall, Angus and Robertson PTY Ltd, London, 1960, p.55.
17. AVIA 22/2454 *Thermite: experiments on use for opening and emptying unexploded bombs*, The National Archives.
18. BBC People's War Article ID A8934672.
19. *Ibid*, ID A8436053.
20. *Ibid*, ID A2025695.
21. Interview Audio Recording 30421, Imperial War Museum, London.
22. Sussex Police Authority records, *Ibid*.
23. *Ibid*.
24. http://www.lettele.nl/?page=nieuwsbericht&id=283.
25. *Ibid*.
26. http://www.bosschebabbels.nl/index.php?option=com_content&view=article&id=65:het-mysterie-van-de-ijzeren-vrouw&catid=37:stadslegendes&Itemid=55.

Epilogue
1. http://news.bbc.co.uk/1/hi/sci/tech/3634212.stm.
2. http://www.atlantic-times.com/archive_detail.php?recordID=1360.
3. http://www.information-britain.co.uk/m/famdates.php?id=77.
4. *The Flying Bomb*, Richard Anthony Young, Ian Allan Ltd, London, 1978, p.152.
5. http://mcfisher.0catch.com/scratch/v1/v1-0.htm and WO 219/2167 and V-1 Flying Bomb 1942–52, Steven J. Zaloga, Osprey, Oxford, 2005, p.39.
6. WO 219/2167 *Trials and proposed use of JB-2*, The National Archives, UK.
7. *Ibid*.
8. http://www.astronautix.com/lvs/loon.htm and http://www.waymarking.com/waymarks/WMB7D3_Wendover_Willie_Replica_V1_JB_2_Buzz_Bomb_Hill_Air_Force_Base_Museum.
9. *Air Launched Doodlebugs*, Peter J.C. Smith, Pen & Sword Aviation, Barnsley, 2006, Appendix XIV.
10. http://space.about.com/od/rocketrybiographies/a/vonbraunbio.htm.
11. *Independent*, obituary, 8 January 2008.

Appendix 1

George Medal Awards (V1 incidents)

Name: T/Major John Pilkington **Hudson** MBE, GM
Unit: Royal Engineers, HQ, Dir of Bomb Disposal
Date of incident: 24/06/1944 – 02/07/1944
Location: Strawberry Hill Farm
London Gazette entry: 15/09/1944, p. 4253

Notes
This was actually an award of a bar to the George Medal that John Hudson had already been awarded the previous year.

John Hudson in 2000.
(*Author collection*)

Name: John Arthur Thorpe **Dawson** BSc, PhD
Unit: Experimental Officer, Ministry of Supply
Date of incident: 24/06/1944 – 02/07/1944
Location: Strawberry Hill Farm
London Gazette entry: 26/09/1944, p. 4438

Notes
Dawson had studied physics at Owens College, Manchester and graduated in 1933. His experience with X-rays led him after the war to write papers such as 'Radon. Its Properties and Preparation for Industrial Radiography'. He was also involved in studying the effects of radiation with regard to the atomic bomb tests in Australia during the 1950s. John Dawson's George Medal was sold at auction in 1993 for £900.

Name: Robert **Hurst** MSc
Unit: Experimental Officer, Ministry of Supply
Date of incident: 24/06/1944 – 02/07/1944
Location: Strawberry Hill Farm
London Gazette entry: 26/09/1944, p. 4438

Notes

After the war Robert Hurst joined the Atomic Energy Research Establishment at Harwell. He worked as a chemist there before becoming the head of a project team studying the potential of various types of nuclear reactor. In 1958 he became the first director of the new experimental fast breeder reactor complex at Dounreay. Then in the early 1960s he became director of the British Ship Research Association. He was awarded a CBE for his services to industry in 1973. He sat on the technical committee of the RNLI for many years and in later life helped out writing up launch reports for the RNLI at Poole. On 16 May 1996, Robert Hurst passed away aged 81.

Name: Lieutenant Eric Wilfred **Sivil**
Unit: Royal Engineers, 20 Bomb Disposal Company
Date of incident: 28/07/1944
Location: Southborough, Kent
London Gazette entry: 02/02/1945, p.685

Notes

Eric Sivil attended the King Edward VII School in Sheffield between 1928 and 1936. A keen sportsman most of his life, he passed away in January 2006.

Eric Sivil in 1955 (*King Edward VII School, Sheffield*)

Name: Leading Seaman Robert **Gribben**
Unit: Naval Party 1572
Date of incident: 15/03/1945
Location: Royal Sluice, Antwerp Docks, Belgium
London Gazette entry: 30/04/1946, p.2086

Notes

The *Glasgow Herald* of 1 May 1946 reported that Scots seaman, Robert Gribben of 128 French Street, Dalmarnock, Glasgow, was awarded the George Medal for great courage, initiative, and devotion to duty in recovering an unexploded V1 from the Royal Sluice at Antwerp. The report stated that 'this difficult operation was finally accomplished after many hours of strenuous work'.

In January 1952, having attained the rank of Petty Officer, he was also Mentioned in Despatches in recognition of Operational Minesweeping and Bomb Disposal services.

Appendix 2

V1 Disposal Instructions

D.B.D. TECHNICAL INSTRUCTION No. 228

Subject:-

        THE GERMAN FLYING BOMB

Distribution:-   D.B.D. List B.
                   R.E. Units holding B.D. Equipment.

Cancellation:- D.B.D. Teleprinter No.45,
                  B.D. Bulletins Nos. 74 and 92,
                  and D.B.D. (Provisional) Technical
                  Instruction No.152 (with Amdts. 1
                  and 2) are <u>cancelled</u> and all copies
                  should be destroyed.

20th January, 1945.

THE GERMAN FLYING BOMB

The information given below is to assist in the identification and disposal of unexploded flying bombs.

1. **General description**

   i. A diagram of the flying bomb ("FZG 76" in German nomenclature) is given in Fig. 1. The body, wings and jet propulsion unit are made of sheet steel but the nose fairing, the elevators and the rudder are made of aluminium alloy.

   ii. Flying bombs are designed to explode on impact but a proportion are fitted with a clockwork long-delay fuze designed to explode the bomb some little time after impact should the impact fuzes fail.

2. **Data for reconnaissance**

   i. Aerial bursts. For various reasons flying bombs sometimes explode in the air, in which case, of course, no crater is formed. Widespread blast effects may be produced and pieces of fuselage, wings and the propulsion unit may be found widely scattered. The latter is composed of an empty sheet steel tube, 11 ft. 3 ins. long and 1 ft. 10¾ ins. in diameter, open at one end and closed by a grill (similar in general appearance to the radiator of a motor car) at the other. The propulsion unit, which becomes intensely hot during operation, is usually crumpled when found and signs of scorching are frequently in evidence around the place where it comes to rest.

   ii. Explosion on impact. When a flying bomb explodes on impact it usually produces a blast effect comparable with that of the larger size parachute mine. When it falls in open ground it forms a shallow saucer-shaped depression, in the midst of which is a central crater 3 ft. to 4 ft. deep and up to about 6 ft. in diameter. Cases have however, occurred where the bomb evidently penetrated the ground before exploding and where the crater has been as much as 10 ft. deep and about 40 ft. in diameter. The propulsion unit, parts of light metal sheeting, pieces of wing or wing spars, tail unit, and control rods may be found in or about the crater. Other parts which may be discovered amongst the wreckage include small electrical units, insulated wire, gyroscopes, metal tanks, lengths of metal tubing and possibly small electric motors and parts of clockwork mechanism. Parts of a balloon cable-cutter, in the form of chisel-edged steel strips, may be found in the wreckage of the wings. Radio parts may also be found, including up to 450 ft. of stranded galvanised wire, either loose or wound on a tubular cardboard former. A number of flying bombs have carried leaflets, whilst others have

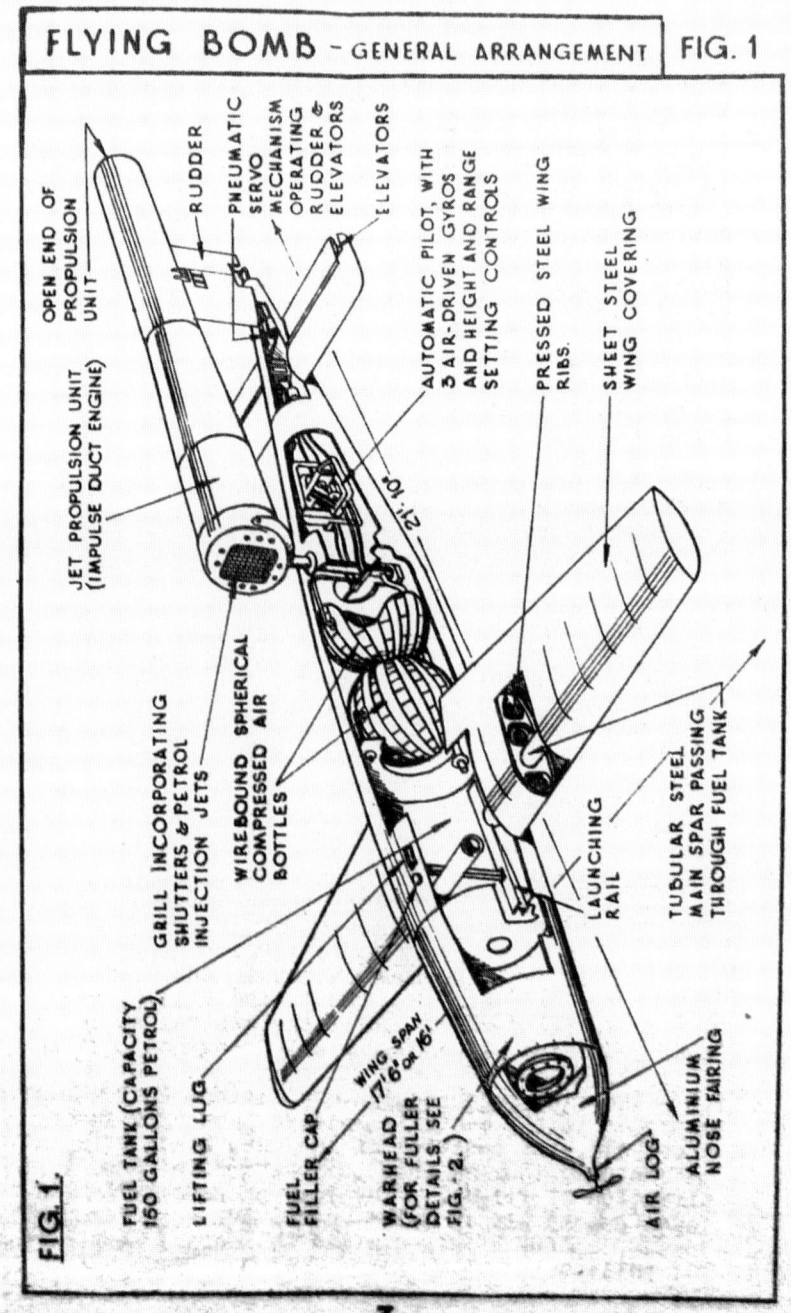

carried up to twenty-three 1 kg. incendiary bombs. Both leaflets and incendiary bombs are scattered by the explosion.

iii. <u>Unexploded flying bombs</u>. Only about 0.02% of flying bombs which have reached the U.K. have failed to explode on impact. U.X. flying bombs may cause considerable damage to buildings etc. which they hit, whilst signs of burning may also be found, caused by ignition of petrol left in the fuel tank. Damage caused in this way is, however, readily distinguishable from the widespread blast effect of an exploded flying bomb.

In more than half the U.X. flying bombs the warhead breaks up on impact, explosive and fuze pockets being scattered up to 400 yards from the point of impact.

The fuze pockets, of which there are three in each bomb, are potentially dangerous and wherever possible should be located and dealt with. If any fuze pocket has a long-delay fuze it may explode where it finally lands, forming a small crater.

NOTE 1. As a result of attack the warhead sometimes comes away from the rest of the flying bomb, in the air. In such cases it will be clear from the condition of the fuselage, that explosion of a warhead adjacent to it has not taken place. It cannot, however, be concluded that the warhead will be unexploded since it may come to earth a considerable distance from the fuselage and will then explode in the majority of cases.

3. <u>Bomb and nose fairing</u> ("warhead") (Figs. 1 and 2)
i. <u>The bomb case</u> is of welded sheet steel, about 0.05 ins. thick, with two longitudinal side seams; a dished rear plate; and a dished front plate provided with a central circular filler hole which is closed by a filling plate, 12½ ins. in diameter, secured by twelve bolts. The bomb is secured to the fuselage by four bolts which pass through lugs on the bomb and petrol tank, respectively.

ii. A central steel <u>exploder tube</u>, partly filled with T.N.T. pellets, is embedded in the main filling. A <u>nose fuze pocket</u>, attached to the filling plate and occupying the forward end of the central exploder tube, contains the normal picric ring and pellets (Fig.7).

iii. Two <u>side fuze pockets</u> are fitted, each normally attached to the central exploder tube by a sheet steel "T", the position of the fuze heads being at 11-o'clock, looking from the rear of the bomb in the direction of flight. The fuze pockets have been found spaced either 2 ins. apart or, more frequently, 14 ins. apart. They contain the normal picric rings and pellets.

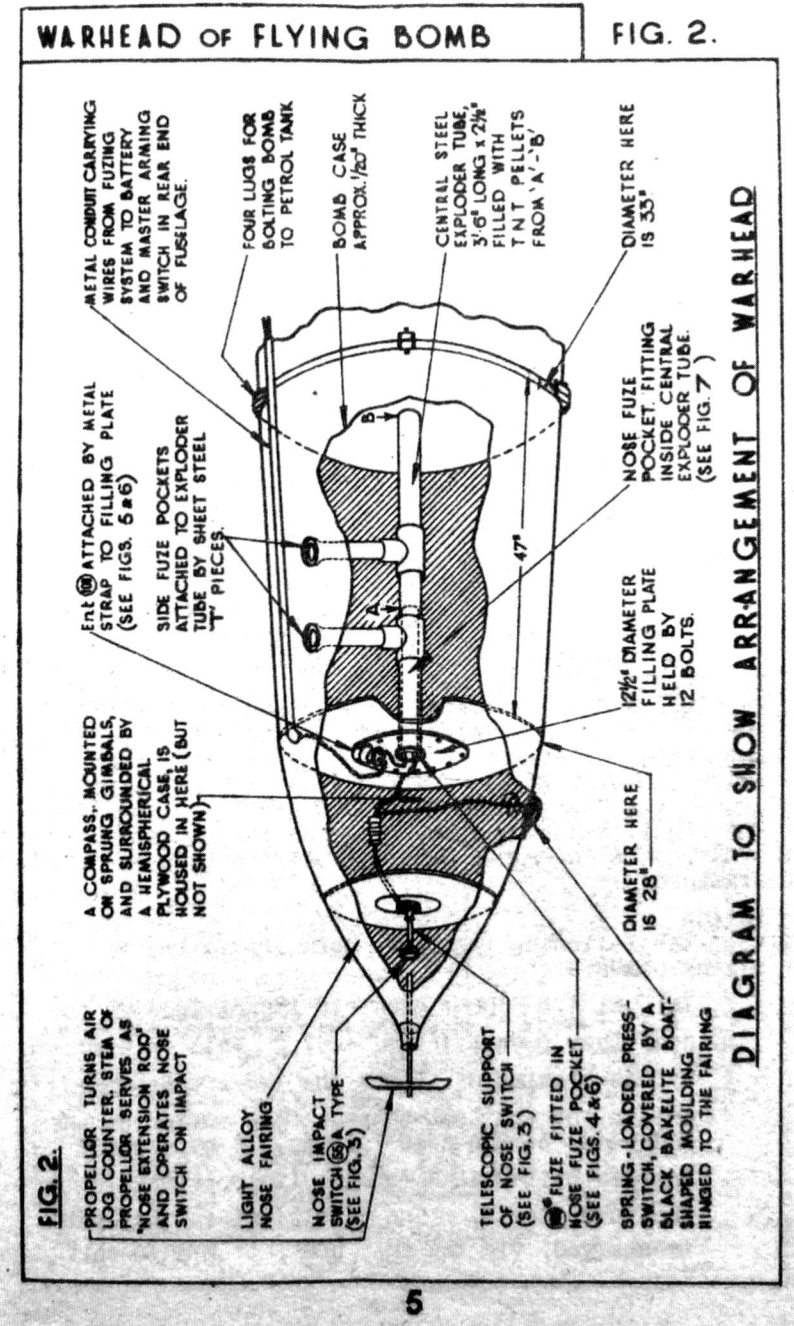

iv. Two types of __main filling__ have so far been found:-
  (a) "52A" - a yellowish mass in which small lumps of white explosive are embedded in a matrix of yellow explosive. It can be readily steamed, but not washed out. (In one example the warhead contained 1,870 lbs. of "52A" explosive).
  (b) "Trialen" - dark grey in colour, which can be steamed out in the usual way.

v. A __nose fairing__ of thin aluminium sheet, is secured to the bomb. A small propellor projects from its forward end, and a black bakelite boat-shaped switch, about 8 ins. long, projects from the underside. Inside the fairing are found a (55)A type impact nose switch, mounted on a short telescopic support (Fig.3); a compass, surrounded by a hemispherical plywood cover; a fuze-shaped object, marked "Ent (106)", mounted on the filling plate of the bomb; and electric wiring. In all cases so far examined the nose fairing was found torn away from the bomb, badly crumpled, and with the internal wiring severed.

vi. __A conduit__ to take the electric wiring etc., is attached to the outside of the flying bomb near its top and runs longitudinally along the warhead to the rear end of the fuselage.

vii. __Markings etc.__ The flying bombs (including warheads) examined have been painted various colours (light blue, dark grey, black); have sometimes shown a red priming coat; and have sometimes been unpainted. The marking "FZG 76" has sometimes appeared on the warhead in large letters. Two bombs (both of which were filled with trialen) were marked with a large red cross, extending round the body of the bomb; two pairs of short white bars; and red and white rings around the rear and forward fuze pockets respectively. The significance of these markings is not known. The cross, bars and rings were not painted on other bombs examined.

4. Fuzing
i. The following fuzes are normally fitted to flying bombs:-

  (a) __El A Z (106)__ * electric impact fuze in the nose fuze pocket (Figs. 4-6). This is actuated by the closing of either the (55)A type nose switch, or the boat-shaped press switch on the underside of the nose fairing, or by a stiff inertia bolt switch (on the (28)B principle) in the fuze itself. The electrical fuzing system is charged, via the Ent (106), from a 30-volt

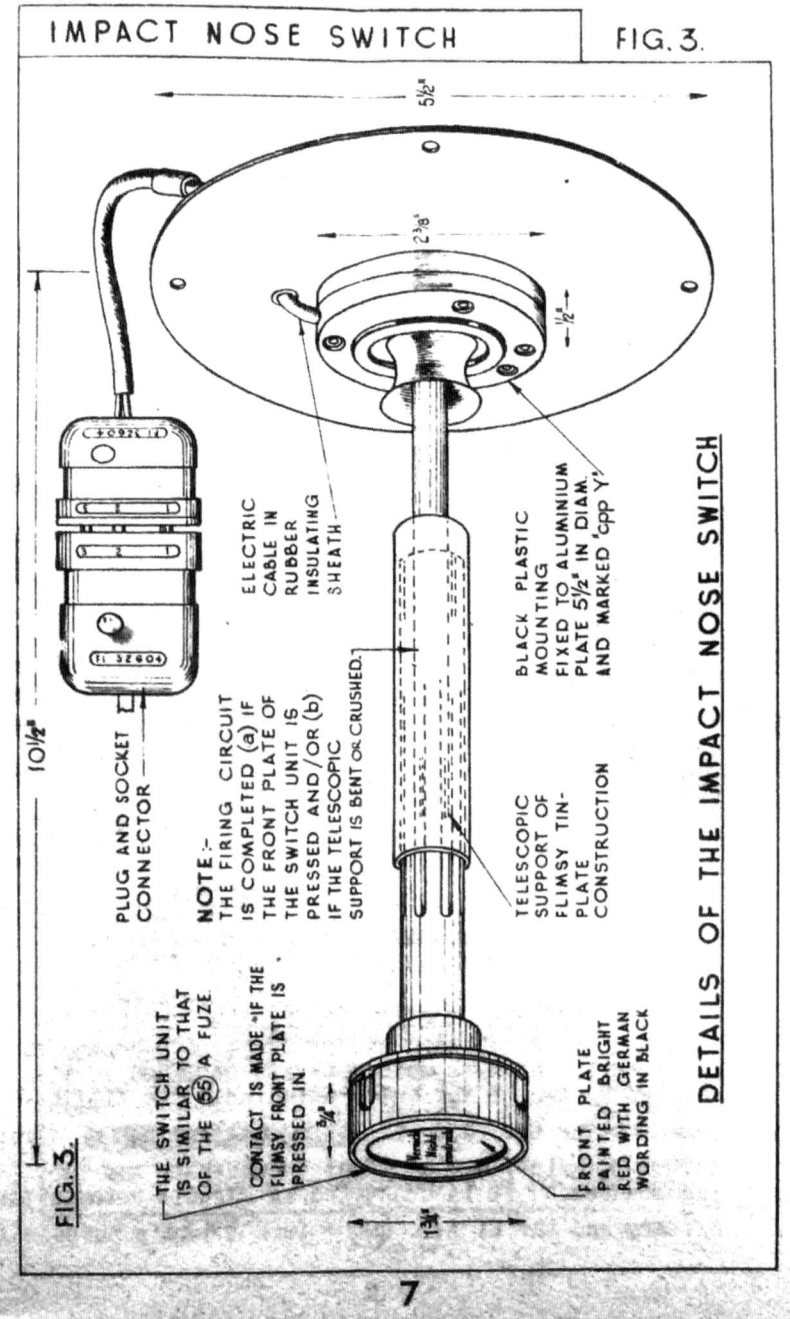

dry battery in the rear of the fuselage, when the bomb has flown about 40 miles (through closing of an arming switch operated by the propellor in the nose).
(b) 80 A allways mechanical impact fuzes, fitted in each of the two side fuze pockets.
(c) 17 Bm clockwork long-delay fuze, which is sometimes fitted in place of one of the 80 A impact fuzes in a side fuze pocket.

NOTE 2. If, for any reason, a fuze has been omitted from one of the three fuze pockets, that pocket will normally be found closed by a brown bakelite storage disc.

ii. No fuze in flying bombs has so far been found fitted with an anti-withdrawal device.

iii. The appearance of the (106)*, 80 A and 17 Bm fuzes, when fitted to bombs, is shown in Figs. 5, 10 and 12 respectively. Relevant details of the fuzes are given in paras. 5 to 7 below.

5. El A Z (106)* electric impact fuze

i. The (106)* fuze contains, essentially, an inertia-bolt switch, an igniter for firing the gaine, and a simple form of arming switch by which the fuze circuit is completed and the fuze is armed when the bomb has flown about 40 miles. The fuze is connected, through an Ent (106) (see below), to a battery in the rear of the fuselage, and is designed to be fired (a) by closing of the (55)A type of nose switch (Fig.3) or (b) if the telescopic support of the (55)A switch is bent or crushed or (c) by closing of the boat-shaped switch under the nose fairing or (d) by operation of the inertia-bolt switch inside the fuze itself.

Since the (106)* fuze contains no internal condensers or battery it is safe when the four wires which normally enter the fuze head are found broken or disconnected.

ii. A fuze-shaped object, marked "Ent (106)," is attached to the forward end of the bomb by means of a steel strip which is welded to its body and is secured by two of the bolts which hold the filling plate. The Ent (106) is not a fuze and has no firing device or gaine, but contains a condenser and two inductances. It is connected in circuit between the battery and the El A Z (106)* fuze and it's purpose is

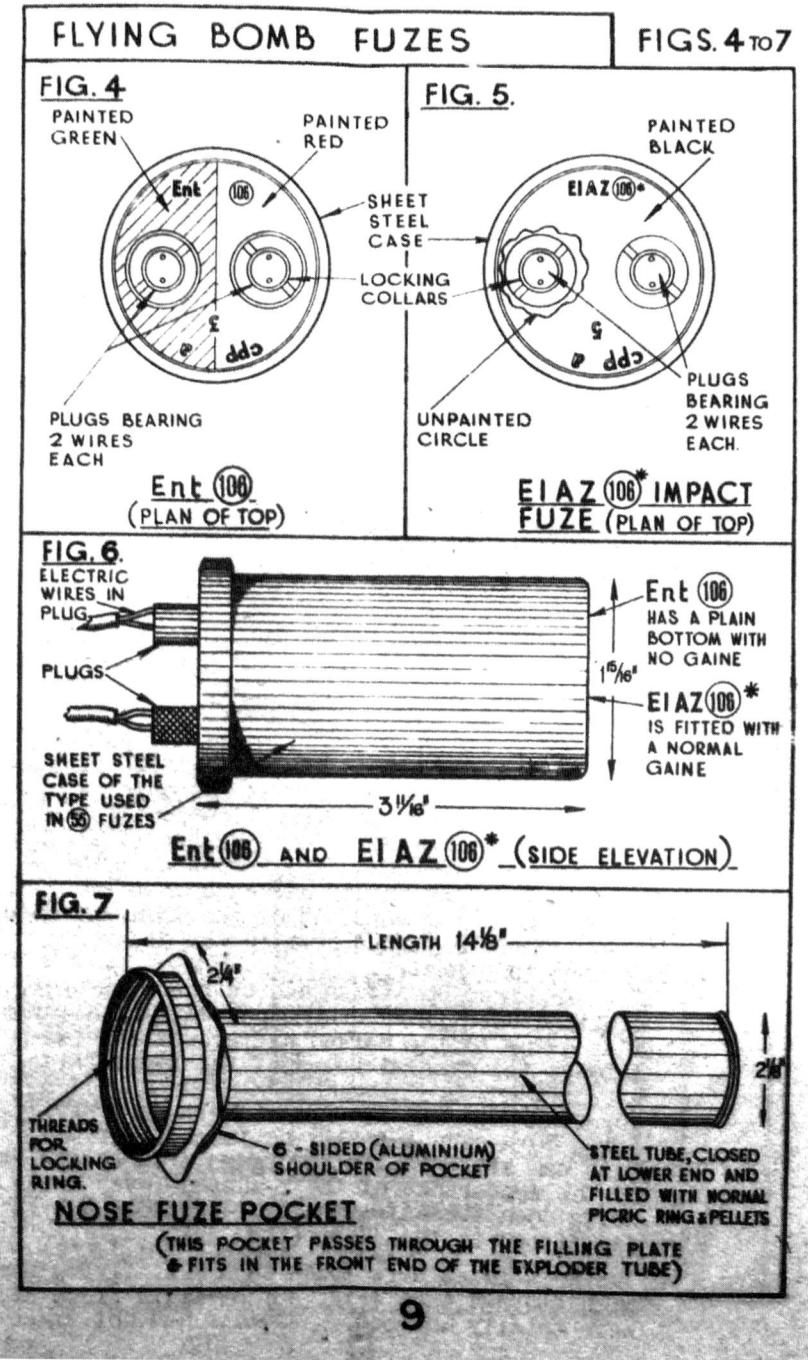

to provide electric energy to operate the (106)* fuze should the connection to the batteries be torn away just prior to impact. The charge is said to leak from the condenser in less than ten minutes after it becomes disconnected from the battery.

6. 80 A allways mechanical impact fuze (Figs. 8-11)
   i. The 80 A fuze has a zinc-alloy case, painted grey except for the top, which is unpainted, but is daubed with a white compound sealing the joints.
   ii. The fuze contains an allways striker unit, consisting of a striker and a cap-holder, each fully floating in a chamber with coned ends. A light creep spring holds the striker away from the cap.
   iii. In an unarmed fuze a hollow spring-loaded safety bolt protrudes through a hole in the cap-holder, locking the striker in a safe position. The safety bolt is held in position by a Y-shaped lever (Fig. 11) which is locked in position by a clockwork mechanism. This is in turn prevented from starting to tick by a flexible spring safety pin, about 1 inch long and 1/16 ins. in diameter, attached to an arming ring.
   iv. A steel wire lanyard, connecting the arming ring to the launching gear, pulls out the safety pin as the bomb is launched. This starts an arming clock, which has a factory-set delay of about 8 minutes.
   v. At the end of this time the clock moves the Y-shaped lever clear of the top of the spring-loaded safety bolt, which is expelled from the fuze. The fuze is then fully armed. On impact, in whatever direction, the striker and cap move towards one another and the percussion cap is pierced.
   vi. For disposal purposes it is necessary to distinguish three conditions in which the 80A fuze may be found, viz:- "unarmed" (arming ring present); "partly armed" (arming ring absent but safety bolt still present, held in the fully home position by the end of the Y-shaped lever); and "armed" (safety bolt absent). These conditions are illustrated in Figs. 10 and 11. It should be noted that the arming ring, on being torn out of the thin aluminium diaphragm, leaves a jagged hole in the diaphragm about ½ inch in diameter.
      (a) An armed 80 A fuze can be readily distinguished from an armed 17 Bm long-delay fuze owing to the marked difference in size between the central holes of the fuzes (Figs. 10 and 12).
      (b) The size of the central hole in a partly armed 80 A cannot easily be judged, but the fuze can still be distinguished from the 17 Bm by the appearance of the Y-shaped lever which holds down the safety bolt of the 80 A but is not fitted to the 17 Bm (Fig. 11).

NOTE 3. Two 80 A fuzes were found in a U.X. flying bomb with storage caps (Fig. 15) still in place and were totally unarmed. Another had not armed

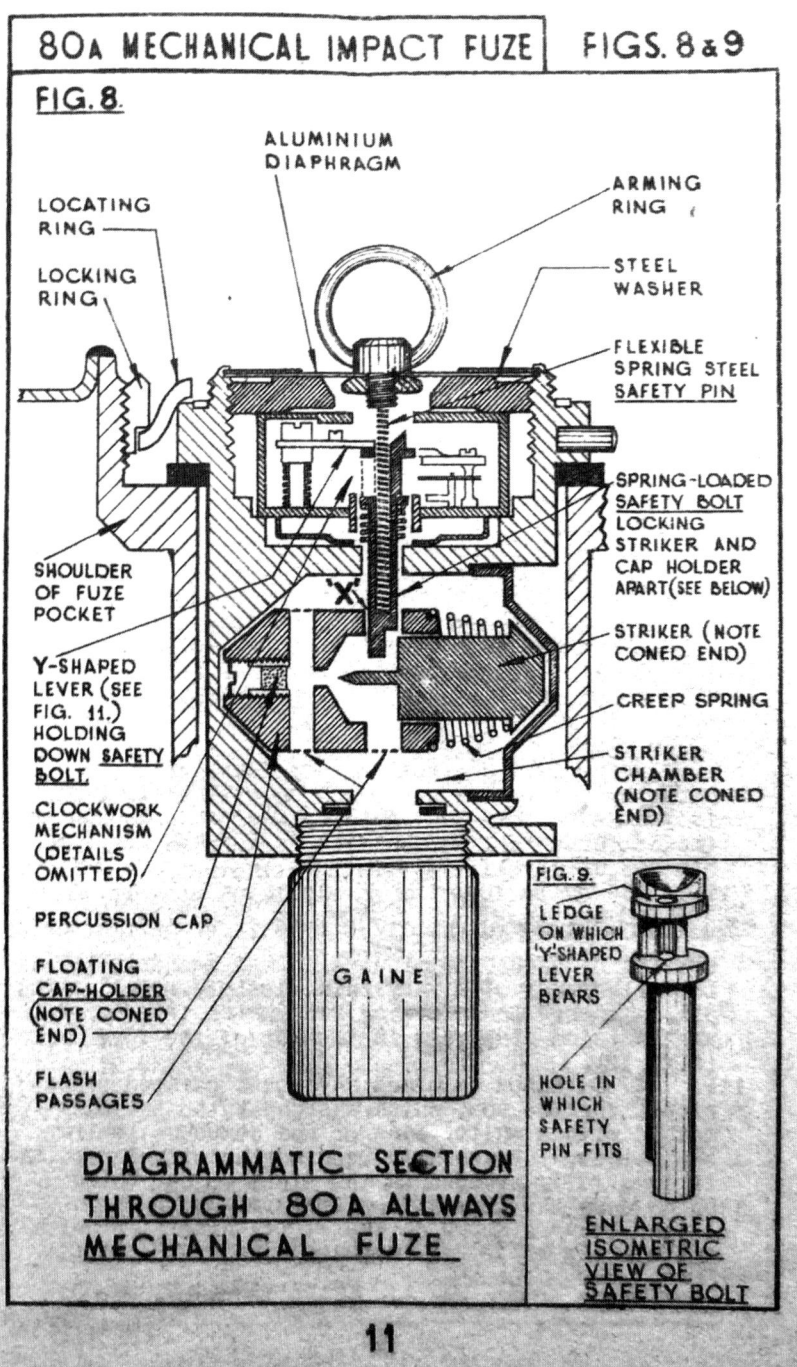

because, although the arming ring had been pulled away, the safety pin had broken off short and so remained in the fuze. In this case the clock could not start, the safety bolt was of course still in place, and its top could be seen blocking the 11/32 ins. hole in the top of the fuze.

NOTE 4. The arming clock of the 80 A fuze is so constructed that the clock may still continue to tick after the safety bolt has been released and expelled and the fuze is fully armed. The clock normally continues ticking until the main spring has run completely down. It is however possible for the clock to stop ticking either before or after the safety bolt has been expelled and yet to restart should it be disturbed (if, for example, an attempt is made to clean out the central hole of the fuze). If the clock restarts it will probably continue to tick for several minutes and, in cases where the fuze was not already fully armed, the safety bolt may be ejected during this time.

Although ticking of an 80 A fuze cannot cause the fuze to operate and can only, at worst, complete its arming, it may not be known which of the two side fuzes is in fact ticking and whether or not the other side fuze is an 80 A. It will be advisable, therefore, for all concerned to retire from the bomb at once if ticking is heard, and to delay return to the bomb until ticking ceases (see NOTE 5).

7. **17 Bm clockwork long-delay fuze**
i. The object of fitting this fuze, which is normally set to function very shortly after impact, is to ensure that the bomb explodes even though the impact fuzes fail to function (e.g. when the bomb makes a comparatively "gentle" landing).

ii. The 17 Bm fuze has an aluminium case in which is fitted a clock of the (17)B type (i.e. maximum delay setting of about 130 minutes) armed mechanically by the withdrawal of a very thin flexible steel arming pin, attached to an arming ring which is secured to an aluminium diaphragm in the top of the fuze (Fig. 12).

iii. 17 Bm fuzes examined have been painted red, except for the top, which was unpainted but was daubed with a white, pink or red compound sealing the joints. Fuzes have been marked either with the figures "17", stamped on the steel washer which holds down the aluminium diaphragm; with the letters "Bm" stamped on the steel washer; or with the marking "Z 17 Bm" impressed on the top of the

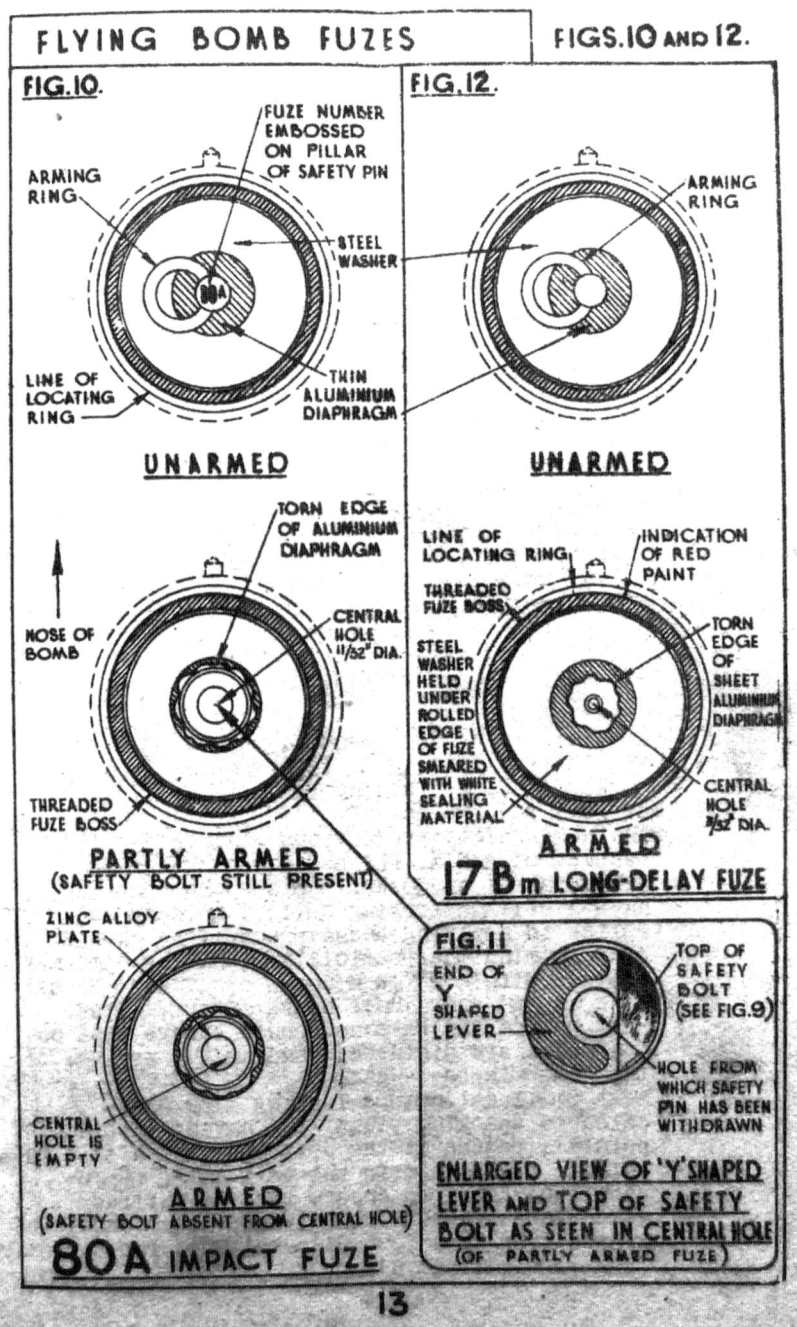

pillar of the arming device with a rubber stamp. Other 17 Bm fuzes were completely unmarked.

iv. When fitted to bombs it may not be possible to distinguish between an unarmed 17 Bm and an unarmed 80 A impact fuze.

When armed, the fuze can be identified (a) by the appearance of the small central hole (Fig. 12), which is 3/32-ins. in diameter as compared with the 11/32-ins. hole in the top of the armed 80 A, (b) by the fact that the fuze case is made of aluminium and (c) by the traces of red paint which may be visible on the side of the fuze near the locating ring.

v. One 17 Bm fuze found in a U.X. flying bomb and subsequently examined was armed, but the clock was not ticking and proved never to have started to tick. It was set to function 32 minutes after arming.

NOTE 5. The 17 Bm has a rhythmic tick similar to that of a pocket watch or a 17 type fuze. The sound of its ticking can be distinguished clearly from that made by the arming clock of the 80 A fuze (see NOTE 4) since the latter is much more rapid, resembling a "whirr" rather than a "tick" when heard through the electric stethoscope.

8. Locking and locating rings

The locking and locating rings are lighter than the type normally used in German bombs and appear to be pressings (note appearance in section shown in Fig. 8). The locking rings have been found distorted and in some cases impossible to remove with a fuze key, in which case it was found possible to remove them by drilling through the locking ring in several places and then springing off the separated sections (see NOTE 7).

9. Safety precautions

i. The Ministry of Home Security has issued the following instructions to C.D. personnel, police etc. in the U.K., in C.D. Training Pamphlet No.1 (2nd Edition) Add. No.1:-

"The same safety precautions will be applied in respect of a U.X. FLY as for an "unburied" H.E. bomb of 1,800 kg. Since the bomb may have a self-destroying device, wardens and Police in applying these precautions should bear in mind the possibility that the bomb may explode within a few minutes or even up to as long as two and a half hours after falling.

Police and Wardens should ensure that no fragments are disturbed pending the arrival of the Bomb Disposal Unit.

Small detonators looking like grey-coloured sparking plugs, but having in place of points a recess at the bottom of which is a red pellet, are fitted to the tail unit. Care should be taken to ensure that no unfired detonators are sent to salvage; they should be handed over to a Bomb Disposal Unit for disposal" (See Figs. 13 and 14).

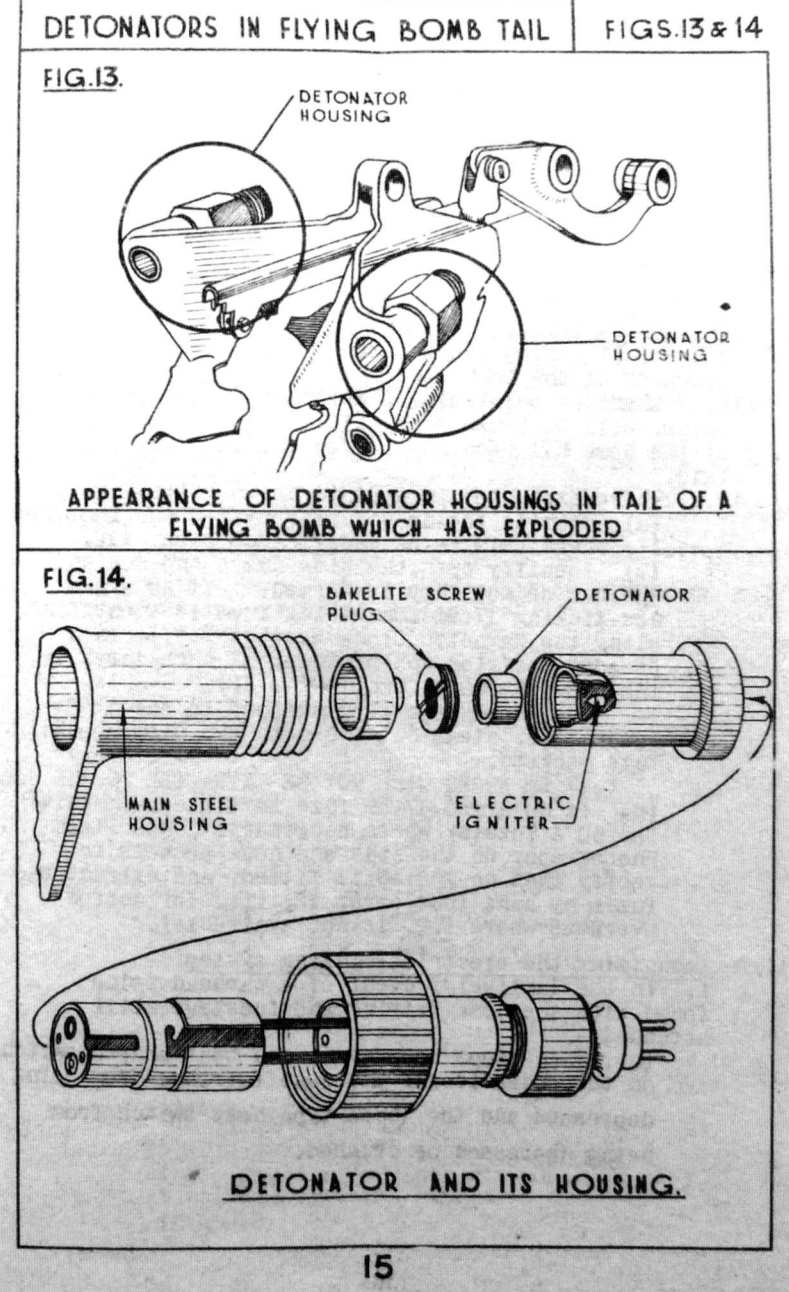

11. As the 17 Bm long-delay fuze has a maximum time setting of about 130 minutes from the time of arming (i.e. normally the moment of launching) there is no point in waiting 96 hours before work is started on Category B-D flying bombs. Work may start 2½ hours after a flying bomb falls.

10. Disposal

The normal sequence of disposal of more or less complete warheads is summarized below. Details of the methods used to immunize the various fuzes are given in paras. 7 to 9.

i. The bomb must not be moved unless and until the necessary steps have been taken to immunize all the fuzes fitted.

ii. If the warhead is fitted with a 17 Bm fuze, or if all the fuze(s) cannot at once be identified, two electric stethoscopes should be applied to the bomb as soon as practicable, one arranged for remote control listening and the other for use by the operator at the bomb (see NOTES 4 and 5).

iii. Whenever permissible, warheads, complete with fuzes, will be blown up in situ as soon as sufficient of the bomb has been exposed to enable a charge to be laid.

iv. Where explosion in situ is not permissible:-
(a) Examine the nose fuzing system and immunize it, where this is necessary (see para. 11).
(b) Identify both the side fuzes and decide whether or not they are armed. If an armed non-ticking 17 Bm long-delay fuze is identified, sling the Magnetic Clock Stopper Des.II in a "ready" position and proceed, if necessary, to immunize the 80 A fuze (para. 12).
Then apply the clockstopper to the 17 Bm (para. 13), steam out the bomb and blow up the fuze pockets.
17 Bm FUZES WILL NOT BE EXTRACTED IN THE U.K.
(c) If no armed 17 Bm fuze is fitted, immunize the 80 A fuze(s) where necessary. Use Field Photography on the side and nose pockets to verify that no ZUS 40 is fitted, and extract the fuzes by hand (see para. 12, iii, for action overseas where F.P. is not available).

11. Immunizing the electrical fuzing system

i. In the (unlikely) event of a warhead being found with the nose fairing and fuselage still attached:-
(a) Take steps to prevent the boat-shaped switch, on the underside of the nose fairing, from being depressed and the (55)A type nose switch from being depressed or crushed.

(b) Disengage the electric cable from the conduit on the top of the bomb and carefully remove the outer sheathing from a part of the cable, taking care to avoid piercing the insulation of any one of the enclosed insulated leads during the process.

Next cut each individual lead, <u>one at a time</u>, insulating the two bare ends resulting from each cut with insulating tape before proceeding to cut the next lead. Then wait ten minutes.

(c) Cut an opening in the thin aluminium nose fairing with tin snips. Then cut (one at a time) the two leads connecting the Ent (106) to the El A Z (106) * and insulate both ends of each lead at once before cutting the next one. It is of vital importance that only one lead should be cut at a time, and that bare ends of separate wires should not be allowed to touch. Fuze El A Z (106) * is then safe.

11. In all cases of U.X. flying bombs so far reported the bomb was found detached from the fuselage and the nose fairing, with the leads severed. In such cases the electrical fuzing system can be regarded as safe <u>ten minutes after the bomb</u> 20 fell, but, where necessary, the leads connecting the Ent (106) and El A Z (106) * should be cut, with the precautions described above.

12. Immunizing the 80 A impact fuze(s)
i. In dealing with 80 A fuzes the following special points should be borne in mind:-
(a) The 80 A fuze is extremely sensitive to shock in any direction. This is evident from the very sharply-pointed striker, the weak creep spring and the fact that the detonator cap is provided with a celluloid top, which probably facilitates entry of the striker.

<u>It is dangerous to move an armed 80 A fuze before it has been immunized.</u>
(b) The striker and cap-holder are both "floating" in a chamber with conical ends. After ejection of the safety bolt the cap-holder frequently rolls round on its longitudinal axis, in which case the safety bolt hole ('X' in Fig.8) will NOT be visible.

If pressure is brought to bear on the cap-holder the striker will slide towards, and pierce, the cap, which will fire.

<u>It is dangerous to clean out the central hole of the 80 A fuze by poking down it with any kind of tool</u> since this may cause the striker to pierce and fire the cap.

11. **In the U.K.** a careful inspection should be made of the fuze, without first moving or disturbing the bomb. Depending on the condition of the fuze the following disposal measures should then be taken:-
   (a) Verify by use of Field Photography that no ZUS 40 is fitted.
   (b) If the fuze is <u>unarmed</u> (i.e. arming ring is present and intact) it can be removed from the bomb by hand.
   (c) If the fuze is <u>partly armed</u> (i.e. safety bolt is present and held in the fully home position by the Y-shaped lever) it should be immunized with the 'J' Equipment, using a threaded adaptor designed to screw on to the fuze head. "Jamming liquid" should be introduced using both vacuum and pressure as described in Technical Instruction No. 220. Then extract the fuze by hand. (Suitable threaded adaptors (Fig. 16) have been issued to all holders of 'J' Equipment).
   (d) If the fuze is <u>armed</u> (i.e. safety bolt is absent from the central hole) but the central passage of the fuze is free from mud etc., the fuze should be immunized with the "J" Equipment, as above.
   If the central passage of the fuze is choked with dirt it will be necessary to clear it before introducing "jamming liquid", to ensure access of the liquid to the striker mechanism.
   Wash out the mud <u>carefully</u> with a fine jet of water (e.g. using the 'J' reservoir as a container and a self-tapping spigot as a nozzle), and then apply "jamming liquid" and extract the fuze by hand. <u>Do NOT attempt to clean out the central hole with any form of spike, wire etc.</u> (see 1, (b) above and NOTE 6).

111. **Outside the U.K.** (where F.P. is not available) disposal should be as above except that the fuze should, where practicable, be extracted by remote control.
   Where the fuze pocket is so distorted that remote control extraction is not possible, the risk of extracting the fuze by hand must be balanced against that of transporting the fuzed bomb to a site where it can be destroyed.

NOTE 6. If a fuze is seen to contain any water, either when found or after being washed out, steps should be taken to extract as much of the water as possible before "jamming liquid" is introduced. It may be possible to suck out most of the water, using a self-tapping spigot and the vacuum pump of the Stevens Stopper, with a short length of rubber cycle valve tubing on the end of the spigot to serve as an "extension", to be carefully introduced as far as possible into the fuze.

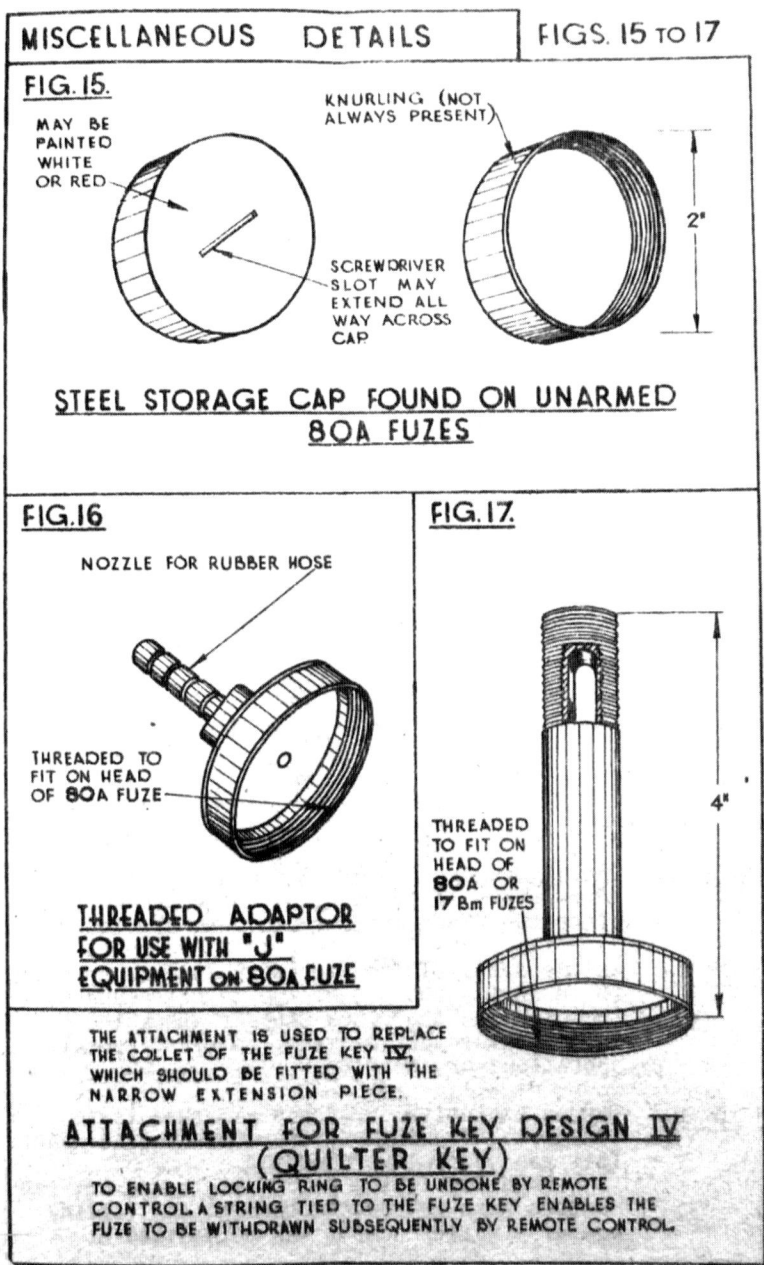

NOTE 7. It may be necessary to drill deformed locking rings to facilitate removal (para. 8, iv) but this should not be done, if the fuze is armed, until it has been immunized with "jamming liquid".

13. Immunizing the 17 Bm long-delay fuze
(Owing to the relatively short maximum delay of this fuze it is assumed that any 17 Bm to be dealt with in a U.X. flying bomb will be non-ticking, even though armed. A non-ticking clock of this type, is however, just as liable to re-start ticking, if disturbed, as a (17) or (17)A type of clock).

i. The 17 Bm fuze can be held stopped (and if necessary stopped) by the Magnetic Clock Stopper Design II in the usual way, the bomb steamed out, and the fuze pockets blown up (see NOTES 9, 10 and 11).
The 80 A fuze should normally be immunized before applying the M.C.S. II to the 17 Bm, but it may, in some cases, be necessary to apply the Clock Stopper to the long-delay fuze before the 80 A has been jammed (e.g. where the clock starts to tick). (Application of a M.C.S. II to a clockwork fuze in one side pocket will cause movement of the floating striker system of an 80 A fuze fitted in the adjacent side fuze pocket. Experiments show, however, that this movement is most unlikely to fire the 80 A even where the pockets are only two inches apart).

ii. To meet the case, in overseas theatres, where steaming-out plant may not be immediately available, an attachment (Fig. 17) has been designed to enable locking rings to be unscrewed by remote control, and the 17 Bm to be extracted by remote control, with the Fuze Key Design IV (Quilter Key) using a system of cords to operate the key. Attachments have been issued on a limited scale.

NOTE 8. Tests show that the 17 Bm clock is very liable to re-start if the current is switched off after the Clock Stopper has been applied.

NOTE 9. As the delay of the clockwork fuze is so short it is advisable to have the magnet slung over the bomb and the Clock Stopper assembled ready for immediate application, before disposal operations are started.

NOTE 10. German explosive "52A" has a marked toxic effect. Steaming-out operators should wear full gas protective clothing.
When one flying bomb was steamed out two nozzles were introduced through the filling plate. Boiler output was maintained at

50 lbs. per sq. inch and all the filling was removed from the bomb in 6¾ hours. It was found that the central exploder tube, with the T-pieces still fitted, fell out when most of the explosive had been removed.

NOTE 11. The general prohibition on steaming out bombs fitted with fuzes which have been filled with "jamming liquid" may be disregarded in the case of bombs fitted with an 80 A fuze immunized by the 'J' Equipment, since the 80 A striker is not spring-loaded.

14. Disposal - where warhead has broken up on impact
   i. In a high proportion of U.X. flying bombs the warhead has broken up on impact. The fuze pockets have been thrown up to 400 yards from the warhead and yet the side pockets have still been dangerous to handle.
   Where a warhead has broken up on impact search should be continued until each of the three fuze pockets has been found unless it is established that those which cannot be found have exploded (see para. 2, iii).
   ii. It is dangerous to move detached fuze pockets containing armed (or partially armed) 80 A or 17 Bm fuzes. Unless it can be ascertained, without moving the pocket or fuze, that a detached fuze pocket contains an unarmed fuze, such a pocket will normally be blown up where it lies, with suitable sandbag protection where necessary.
   iii. Where pockets are seen to be fitted with armed 17 Bm, 80 A, or unidentified fuzes, and where explosion in situ is undesirable, an attempt might be made to pull such pockets by remote control to the nearest site where they can be blown up (after waiting 2½ hours before re-approaching pockets which are, or may be, fitted with armed 17 Bm fuzes).
   iv. It will normally be necessary to collect and dispose of the main filling, which may be scattered over a wide area. Rubber gloves should be worn by men handling the "52A" type of filling.

D.B.D. Technical Instruction No. 226

ERRATA

1. Page 2.   Underline the main heading and the headings of paras 1 and 2, and the headings of sub-paras 2,i and 2,ii.

2. Page 2.   Sub-para 2,ii <u>add</u> "usually" at end of line 4.

3. Page 4.   Line 4, <u>amend</u> "0.02%" to read "0.2%".

4. Page 10.   <u>Delete</u> underlining in lines 1 and 2.

5. Page 10.   Line 4 <u>amend</u> "ten" to read "20".

6. Page 17.   Lines 11 and 25 <u>amend</u> "ten" to read "20".

7. Page 20.   Line 2, <u>delete</u> "iv" after "para.8".

8. Page 20.   Sub-para 13,i, line 14 amend first word to read "adjacent".

AFTER MAKING THE ABOVE CORRECTIONS PLEASE REMOVE AND DESTROY THIS SHEET.

Appendix 3

V1 Fuze Technical Information

Extract from *Fuze News*

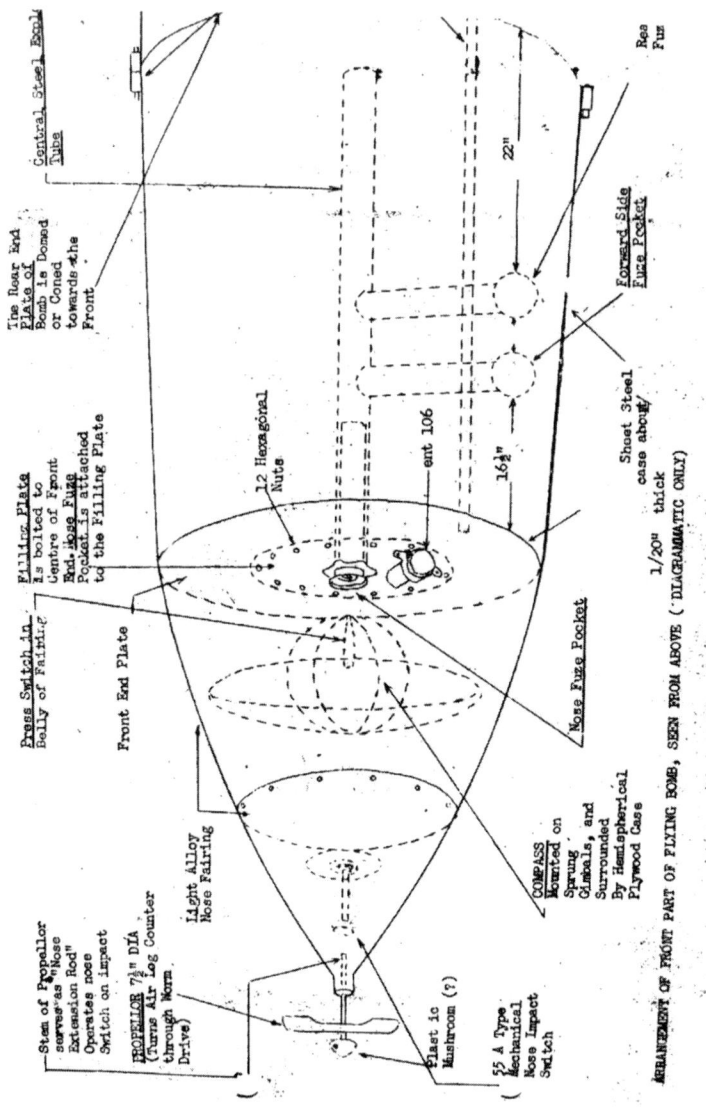

V1 Fuzing Arrangement.

Extract from *Fuze News*

SECRET

SUBJECT: El.A.Z. (106)* German Fuze.

Bomb in which Used: German Robot Bomb V-1

Description: The El.A.Z. (106)* German Fuze is contained in a sheet metal case similar to that used in the (55). The top has two sets of wires plugged in, each plug held by a locking ring staked in place. The fuze is apparently mounted in the forward bulkhead of the warhead with the top of the fuze facing the nose of the bomb. One of the plugs is wired to the diaphragm type switch in the nose of the bomb. The other is wired to the Ent. (106) fuze. We have not as yet seen this fuze, but it apparently contains a condenser of about 1uF capacity which is charged by a battery also contained in the bomb. The wiring diagram of the El.A.Z. (106)* is shown also a diagram of the Pressure Switch.

Operation: After launching the condenser in the Ent (106) is charged from the battery in the bomb. A switch attached to arming vanes in the nose of the bomb completes this circuit.. The charged condenser is thus connected across A and B so that the igniter bridge E is fired, igniting the Thermite Pellet. Heat from this melts the Plastic Tube, which is a part of Spring-Loaded Plunger 3. Plunger moves inward breaking the circuit through Igniter Bridge E. Heat from the burning thermite then melts the Solder Pellet to which Plunger 1 and 2 are attached by wires. When Solder Pellet melts, Plungers move outward. Plunger 2 makes a contact connecting A with igniter bridge F. Plunger 1 breaks a shunt across igniter bridge G. Two firing circuits are now established; from the condensers in Ent (106) to A, to Plunger 2, igniter bridge F, Inertia Switch, to B, and back to condenser; from condenser to A, plunger 2, igniter bridge G, C, Pressure Switch, D, B, and back to condenser. Upon impact, the fuze can be fired by the Inertia Switch or by the Pressure Switch. In the case of the latter it is of interest to note that not only will it fire by the plates being pressed together but also if the nose of the bomb is distorted sufficiently so that the outer tube and inner tube contact each other.

Remarks: Cutting the wires running to the Ent (106) fuze will render the fuze harmless. Cutting the wires running to the Pressure Switch will fire it.

Extract from *Fuze News*

SUBJECT: German Fuze 80A.

Bomb used in: German "Robot" Bomb V-1.

Description: Fuze consists of a die cast body into which is fitted an "allways" type of Striker and Primer Holder similar to that used in the (24)A. They are held apart by a Creep Spring and (in the unarmed state) by a Safety Pin. The Safety Pin, which is spring loaded upward, is held down by the "Striker" of the Arming Clock, which is a (67) Fuze with only slight modifications. The Arming Pin, which is fastened to an Aluminum Disc held in place by a Steel Washer crimped to the fuze body, holds the Stop Arm of the clock. The top of the fuze is covered by a screwed cap.

Operation: The cap is removed from the fuze and the ring in the end of the Arming Pin is fastened to a lanyard, which is pulled at the time of launching. When the Arming Pin is thus removed, the Stop Arm moves over allowing the Clock to start. After about 7 minutes, the "Striker" of the clock released, allowing the spring-loaded Safety Pin to be ejected from the fuze. The fuze is now armed. Upon impact, the Striker and the Primer Holder are forced together, setting off the No. 26 Detonator which in turn sets off the gaine.

Remarks: This fuze is apparently used as a bomb fuze and not as a "destructor" as in the Hs293 Glider Bomb. It is very sensitive, a drop of from 5/8" to 1-1/4" being enough to set it off when armed. If the Arming Pin and Aluminum Disc are still in place, fuze is safe. If "Striker" of clock may still be seen thru hole in top of fuze to be holding the Safety Pin in place, fuze is not armed but may become so if clock should start again. If Safety Pin is missing, fuze is definitely armed.

80A Allways Fuze.

Extract from *Fuze News*

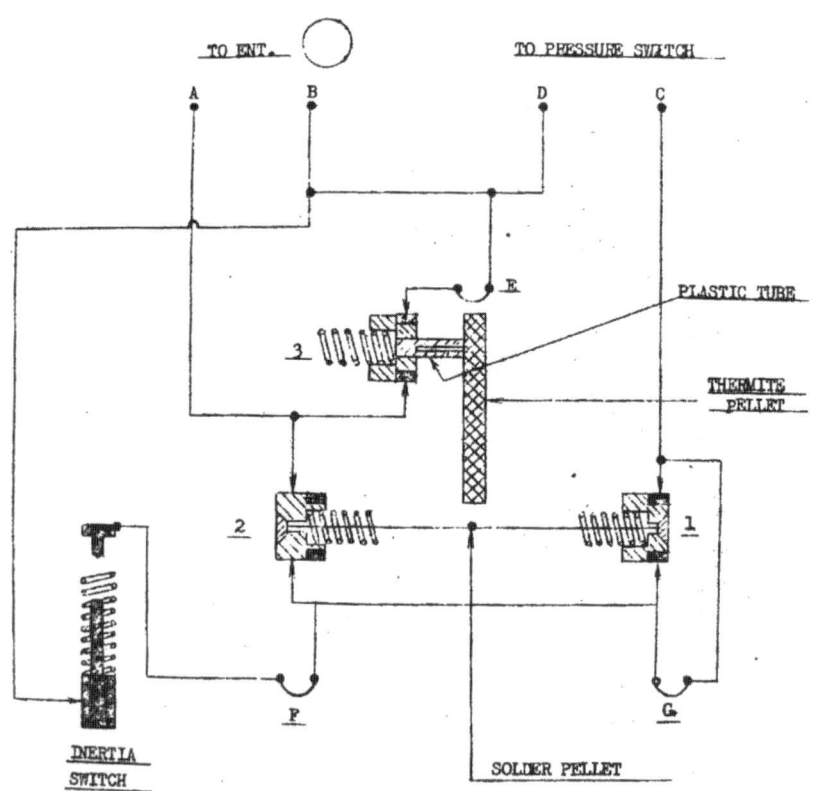

EI.A.Z 106*

Appendix 4

V2 Disposal Instructions and Nose Fuze Technical Information

SECRET

Instructions to R.A.F. Bomb Disposal Units
Air Ministry Instruction No :-1332
Air Ministry Reference No :- S.102396/0.10
BD/S.376/Arm – 69
Date :- 29.7.45.

Nationality :- German
Subject :- Long Range Rocket A4

1. GENERAL
(i) The following instructions are to be read in conjunction with A.M.I. No.1331.

2. DISPOSAL
(i) Expose the rear end of the warhead, cutting away the after body of the rocket as is necessary. Precautions are to be taken to avoid any vibration or movement of the bomb.
(ii) If the Sterg unit is missing, or if <u>all</u> cables leading from it to the nose and rear fuzing components of the warhead are severed, the electrical fuzing system may be considered inoperative.
(iii) Insulate, by means of insulating tape, each exposed end of each severed lead, taking care that the ends of the various leads are not brought into contact with each other.
(iv) If the Sterg unit is present, and the cables leading from it to the nose and rear fuzing components of the warhead are intact, it must be considered that the warhead is "LIVE", and either of the following procedures are to be adopted:-

V2 Disposal Instructions and Nose Fuze Technical Information 249

    (a) Remove the jack plugs from the Sterg unit one at a time and insulate each as it is removed by using insulating tape, or

    (b) Using a sharp knife, shave away the outer sheath of each cable separately to expose the inner insulated leads. Cut each lead separately and insulate the bared ends as soon as cut, with insulating tape.

(v) The warhead portion of the fuzing system is now completely immunized.

(vi) Remove the Sterg unit by unscrewing the two holding nuts.

(vii) Unscrew the three wing nuts holding the rear fuzing component to the filler plate and remove the rear fuzing unit.

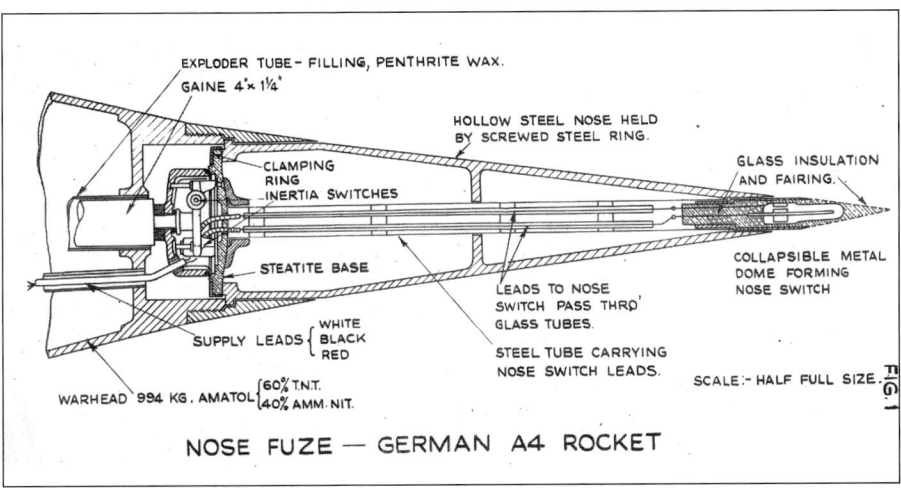

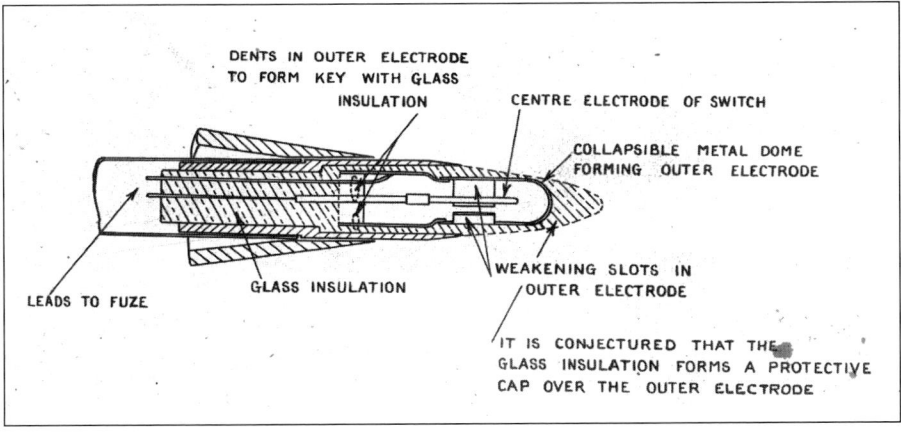

(*The National Archives: ref.AIR 2/9224*)

(viii) Lift out the rear fuzing unit gaine.
(ix) Unscrew the steel locking ring at the nose of the warhead and remove the nose fuzing component.
(x) Remove the nose fuzing component gaine.
(xi) In the event that the warhead is required and has to be stored, seal both nose and tail ends of the central tube, ensuring the seal is moisture proof, before storage.
(xii) In the event that the warhead is to be destroyed, either of the following procedures are to be adopted:-
    (a) Destruction by demolition, where such demolition is acceptable, or
    (b) Remove the warhead to a suitable boiling out site, remove the filler plate and boil out in the normal manner. The explosive content (main filling) is to be subsequently burnt under precautions, and the central exploder tube demolished under full safety precautions.

Air Ministry, 0.10
Distribution List "G.7".

Leonard Harrison
Wing Commander
Head of 0.10.

Index of People, Places and Units

People
Aalpol, O. J. J. H. 123, 170
Allison 115
Anderson 8
Antwerpes, Franz Joseph 176
Ashe, Lincoln F. 5
Ashmeade, Doreen 43, 44

Bainbridge 90
Baldwin, Charles 143
Ballard 19
Barker 62, 64
Barnes, David 22
Bartlett, Roy 199
Bassett, Charles John 11, 59, 60
Bateman, H. H. 7, 26, 31
Bayer, Rolf 172
Beer, E. J. 88, 112
Behrendt 25, 30
Bellamy, David 196
Bens, Peter 176
Beswick 89
Biggs, Alfred John 39
Blackmore 140
Blaney 145
Blyth, R. 131
Boorman 88, 89
Booth 110, 111, 115
Brandreth 57
Briggs 103
Brinton, Cecil 124
Brockbank, Jim 118
Brock 196
Brooks 54
Brown 115
Bube 96
Bullard, Hazel 43, 44
Calvert 13

Carlile, Frank 20, 21
Carter, Edward 42, 68, 71
Cartwright, Charles 87, 111, 112
Castellan, Peter 95
Chantrell, Tom 194, 195
Chapman, Les 51, 52
Christiansen 13, 14
Churchill, Winston 5
Clark 69
Clark, Boyd 90, 91
Clarke, Rosemary 201
Clary, Walter 53
Clinch, Bert 52, 53
Cocks 19
Collins, George W. 98, 101, 162
Comfort, William 9
Commin 28
Connolly 107
Cook 124, 126
Cook, James 26
Copeland, Lawrence 195
Cotton, L. 118, 119
Court, Fred 118
Cox 19
Cox (F/Lt) 107, 110, 111
Coyne, J. P. 118
Craig, William 142
Cripwell 15
Cronyn, Hugh 2
Crouch, Horace 15
Curry 90

Danks 110
Dawson, John 19–21, 24, 25, 28, 31, 34, 220
Day, Reuban 84, 85
Deane, Thomas J. 59, 60, 65
De'ath, G. D. 30, 31, 65
Dequeker, Camiel 94

Devine 15
Dinwoodie 140
Doreleijers, Gabriel 180
Dowden 140
Doyle, Ian 51
Dyer 111, 115
Dykes, A. S. 118
Dykes, D. 126

Earl of Suffolk 151, 196
Edwards (P.O.) 57
Edwards (Sgt) 169
Edwards, Roy 164
Eigell, Robert W. 155, 156
Eisenhower 112
Ella 42
Elliot, Hazel 46
Empett 106, 107
Evans 54
Evans, W. R. 29

Feather, William Anderson 31, 32
Feldman, John E. 24–8, 48, 96, 155
Fenton, A. 169
Fiers, Jan 181
Finney 142
Fleming 103, 110, 169
Frake, G. A. 65
Fullerton, Robert 95

Gaymer, Robin 191
Georges 4
Gerhold 72, 73, 75, 79, 80
Gibbs 136
Goodyear 52
Gray, J. 29
Greatbach 25, 28, 31
Gregory 110
Grew 15
Gribben, Robert 132, 221
Grover 136

Hallows 110
Harderlie, Eugene 3
Hardham, C. R. 53, 54
Harding, Fred 34
Harris, J. E. 118

Harrison 69
Hartley 66
Hauxwell 103
Heath 13
Henderson, Thomas 142
Heppell, Ellen 190
Herridge, Dan 84
Hetterley 52
Hewitt 126
Hillmann, Kurt 133
Hitler, Adolf 3, 5, 11, 19, 196, 201, 205
Hoar, T. P. 29
Hoare 136
Holder, Jack 142
Hollard, Michel Louis 14
Holness, Bob 196
Hood 124
Horan, H. J. 131
Horsfall, G. F. 61, 62, 64, 65
Howard, John 164
Hudson, C. S. 19
Hudson, John Pilkington 15, 17, 19–22, 24, 25, 28–30, 32, 34, 35, 39, 40, 149, 150, 206, 207, 220
Hunt, Herbert James 61, 63, 64, 161
Hurst, Robert 15, 17, 19–22, 24, 25, 28, 30, 32, 34, 35, 220, 221

Jamieson, E. H. 100
Jenkinson, Jim 187
Jensen, G. M. 165
Johnstone, James 142

Kane, Thomas J. 25–8, 96
Karsh 24
Kayser, Eric 170-73
Keeble, Peter 132
Kelch, Russell 91
Ketley, Donald 199
Kilgariff, Danny 192
King George V 162
Kisielewski, Wladyslaw 92
Kleine-Toereers, A. 120
Koch, William 24
Kuiten 181

Ledger, Tom 38
Lenz, Horst 186, 187

Index of People, Places and Units

Lepelaar, Corrie 118
Lessmann, Volker 176, 177
Lewis 169
Lewis, Roger 164
Light 110
Lynn, S. C. 75, 76

McCollum, Leo E. 90
McLardy 15
McWhinnie, J. S. 118
Mackerras D.J. 55
Maries, Raymond 131, 132
Marshall, R. O. St J. 54
Martin (Capt.) 68
Martin (F/Sgt) 103, 106, 111
Martin, John 197
May, K. 195
Mayer, George 176
Medhurst, Alfred 51
Meijers, Antoon 178, 180
Melville 110
Merriman 146
Merrylees 153–6
Middleton 87
Millchamp, Ivy 84
Miller, John 163
Mitzke, Walter 172, 173
Moffatt, Bruce 56
Moon, Ian 190
Moxey, Eric 154
Munford, Thomas 142
Murphy 126

Newitt, D. M. 19
Newitt, Clive 19, 20, 24, 25
Nicholls 42
Norris 69
Nowers, Don 196

O'Connor, J. M. 115
Oram 103
Ouvry, John 8

Patchell, James 116
Payne, Charlie 198
Phemister, J. 6
Phillips 126

Pink 115
Pitcher 59
Polson, A. G. 2
Popper, Karl 34

Queen Mary 162
Quilter, Raymond 32, 153

Rausch, S. F. 101
Reece, Thomas E. 97–100, 116
Rees, W. D. G. 142
Retallick 114
Richards, Brian 119
Richardson, James 51, 52
Ritchie 110
Roache, Thomas J. 99
Robertson, James 140
Robinson, Rex 46
Rubery, Douglas 193
Rubery, Stan 110
Rudd 57
Ruesink, Gordon A. 96
Russell, Arthur 131, 132

Salisbury, John 142
Sayer, H. B. 152
Scamell, Kenneth 103, 106, 107, 110, 164
Schmid, Dietmar 186
Schmitz, Hubert 172, 173
Schneider, Herbert 176, 183
Scott, Ray 44
Sewell, John 194
Sharman, Tom 118, 119
Shorling 26
Simpson, Roland 142
Sivil, Eric 52–4, 221
Slater 124
Smith 24
Smith (Farmer) 30
Smith (Capt.) 169
Soames, Nicholas 138
Stevens, J. C. 150, 152, 163
Streeten, Gilbert 65
Swinson, Charles 79

Taylor 119
Thistleton-Smith, Geoffrey 8,9
Thomas, A. C. 165

Thomas, Wally 158–60
Thomason 188
Thorns 9
Tibben, Anton 200
Tollett, Bob 51
Tuckwell, John 163
Tyson, G. 61

Ulmer, Joachim 172, 173

Van der Sleesen 121
Van der Velde 170
Van Maren 181
Van Sleezen, J. 170
Van Vugt, Herman 200
Van Vugt, Piet 200
Vassie 48
Von Braun, Wernher 205, 206

Wakeling, Eric 134
Walden, Leonard 8, 9
Wareing, G. E. 118
Warner 28
Warner (Col) 128
Wells, H. G. 71
Westbrook 126
Williams, J. 89
Williamson 110
Wiltshire, William 22
Wood, C. R. 88, 89
Woodruff 101

Yard 59–61
Yorke-Smith 124

Places
Aachen 170
Acaster Malbis 187
Addington 58, 196
Alabama 206
Alamogordo 205
Aldeburgh 60
Aldwych 31
Alkmaar 132
Almelo 118, 167, 168, 188
Amiens 93
Amsterdam 92

Antwerp 92, 101, 106, 110, 111, 114, 126, 130–2, 173, 206, 221
Appledoorn 170
Arnhem 95, 118, 182
Arromanches 87
Asch 96
Ashford 196, 198
Ath 106
Australia 220

Backbebo 67
Badminton 162
Bakerloo Underground line 20
Balcombe 55
Barking 163
Battersea 21
Battle 22, 42, 56
Bayeux 87
Beckney Farm 195
Belgium 90, 92, 96, 98, 99, 101–3, 106, 124
Benenden 42
Bercham 111
Bere Farm 8
Berkeley Court 17
Betteldorf 186
Biggin Hill 154
Birmingham 39, 119
Blackheath 81
Blankenheim 170–3
Bleicherode 128
Blizna 67
Boarhunt 8
Bodilsker 13
Bolkshoek 168
Boreham 15
Bornholm 13, 14
Boulogne 42
Bourg Leopold 126
Bradford 187
Brandy Hole 195
Brasschaat 96
Breebrocksweg 200
Bremen 125
Brilon Forest 112, 113
Bristol 207
Bristol Channel 2
Bromma 67

Index of People, Places and Units   255

Bromskirchen 112
Bronsfeld 173
Brook Farm 57
Brookland 38
Brookwood 55
Brosarp 13, 14
Brussels 92, 106, 114, 124
Büchel 115
Bucholz 180
Bungay 61
Burgess Hill 34
Burnham-on-Crouch 82
Butley Abbey 62

Caen 88
Calfven 120
Cambridge 29
Capel Green Farm 62–5
Capel St Andrew 62, 65, 66
Celle 116
Chase Wood 57
Cheam 196
Chediston 83
Chelmsford 15
Chelsea 42
Cheshire 66
Chipping Ongar 69
Chiswick 68
Church Fenton 187
Cologne 96, 115, 176, 184
Corton 60
Corton Sands 164
Cowbeech 71
Crag Pit Farm 60
Crawley 45, 46
Crayford 139, 140
Creasey's Farm 73–5, 79
Croydon 58, 65, 196
Cuxhaven 128, 180

Dagenham 138
Dahnen 107
Dayton 203
De Klijte 94
Dearbusch 115
Delettes 88
Demerstraat 120

Den Helder 182
Dengie Marshes 83
Denmark 13, 133
Deptford 81
Detmold 128
Deventer 118
Dieppe 88, 93
Diksmuide 106
Diss 193
Doel 111
Dounreay 22, 221
Dover 48, 57
Downe 38–41
Downe Court 38
Dunkirk 154
Dusseldorf 172

Ealing 199
Earls Colne 158
East Ham 64, 138
Eastchurch 194
Eckernforde 112
Eckfeld 168, 180, 185
Edmonton 49, 193
Eefde 178
Eggebek 115, 133
Eifel 92, 170, 171
El Paso 205
Elgin Field 203
Elham 57
Elsinore 13
Eltham Green 119
Enfield 139
English Channel 3, 42, 48, 57, 87, 133, 190
Epping Forest 71
Erith 140
Essex 15, 42, 58, 59, 68, 69, 71–4, 79, 80,
    82–4, 158, 166, 188–92, 195
Evere 114

Fairlight 15–19, 21, 24, 28, 29, 36, 48
Fairmead Bottom 71
Falconwood 43
Farnborough 19, 23, 34, 37, 67, 69, 79, 102,
    104, 105
Faygate 42
Felixstowe 65, 84

Feltham 140
Flevo Polder 174
Florennes-Juzaine 99
Florida 203
Fontaine-sous-Preaux 108
Fort Bliss 205
Fort Halstead 19
Foucarmont 93
Foulness 71
France 4, 14, 88, 91, 95, 101, 108, 109, 126, 154
Frant 57
Freeman Field 189
Frenchman's Bay 140

Gelderland 178, 186
Germany 3, 5, 7, 34, 92, 96, 107, 125, 133, 167, 168, 170, 171, 176, 183–6, 203
Gess 93
Ghent 106, 107, 110
Glasgow 221
Gloucestershire 162
Gorssel 179
Gottengen 128, 129
Great Baddow 59
Great Yarmouth 164
Greenwich Park 81
Grimbergen 92
Groenedaellan 103
Groesbeek 118
Grosvenor Square 26
Guestling 199

Haagse Bos 72
Hachenburg 96
Hadlow 55
Hague 71, 72, 80, 169
Halesworth 83
Hallingbury 166
Hamburg 180
Hammer 13
Hampshire 22
Hampstead Heath 20
Hanau 93
Hang Grove Wood 38
Hanover 125
Hänscheid 176, 183

Hanwell 199
Harlow 69
Harrow 15
Harwell 221
Harwich 31, 84, 85, 190
Hastings 28, 29
Hatzfeld 112
Hawkinge 57
Hayes 193
Heathfield 34
Heemstede 94
Heeten 200
Heidelager 5
Hekelingen 181
Hekelingen 177
Hellendoorn 96, 120
Helsingborg 13
Hemel Hempstead 196
Hertogenbosch 200
Heston 140
Hildenborough 56
Hillesheim 93
HMS *Vernon* 8, 9
Hoboken 110
Holland 95, 120, 124, 169, 188
Holloman Air Force Base 203
Holten 118
Holywell Park 139
Hopton 59, 60
Hornchurch 79, 80, 191, 192, 200
Horsham 21
Horsmonden 38
Hull 165, 204
Huntsville 206
Hurstmonceaux 71
Hutton 72–4, 76, 191
Huybergen 120
Hyzinghen 110

Icklesham 48
Ijzeren Vrouw 200
Illinois 24
Indiana 189
Ipswich 139, 191
Isle of Dogs 187
Isle of Sheppey 194

Index of People, Places and Units 257

Japan 203
Jeantes 90
Julianapark 120
Junkerath 94
Juvencourt 101

Kalmak 67
Kaltenkirchen 168
Karlshagen 11
Karlskrona 14
Kassel 128
Keerbergen 96, 101
Kensington 19, 20, 61, 80
Kent 11, 28, 38, 42, 43, 48, 49, 51, 55, 57, 62, 193–8
Kiel 111
Kings Cross 2
Kingsland Farm 195
Klein Bodungen 128, 129
Knivingaryd 67
Knock Wood 49
Knokke 114
Koblenz 115
Kropp 115
Kröslin 190
Kruisland 120
Krummel 168

Le Havre 87
Lea Green 81
Leck 168
Leeds 187
Leiston 59
Lettele 200
Leysdown 194
Leyton 139
Liege 92, 98, 108
Lierre 103, 106
Lissendorf 176
Little Maxfield 199
Little Plumstead 69
Lodge Field 56
Lokeren 124, 126
Lomers 115
Lommersdorf 115, 170
London 2, 4, 14, 16–19, 26, 31, 38, 43, 44, 49, 61, 63, 64, 68, 81, 84, 90, 119, 124, 138, 146, 160, 161, 163, 187, 188, 193, 199, 207
London Bridge 187
Long Island 203
Lower New Barn Farm 15, 16
Lowestoft 22, 59, 60, 136
Luttenberg 120
Lympne 48, 57

Magdalene Laver 69
Magny-en-Vexin 96
Manchester 66, 195, 220
Manderscheid 168
Manor Park 145
Manston 42, 50
Mark Cross 68
Markelo 164, 180
Mataram 95
Merton 49
Middenleane 84
Milzenhauschen 173
Mittelbau-Dora 202
Mittelwerk 127, 129
Moerbeke 106
Mojave Desert 203
Mountnessing 191, 192
Munkzwalm 107
Murnserleane 95
Muroc Dry Lake 203

National Gallery 19
National Physical Lab 8, 157
Nazeing 58, 59
Neaves Farm 61
Netherlands 84, 94, 95, 121–3, 167, 169, 173–8, 181, 182, 200
Nettersheim 170
Neuludwigsdorf 112
Neuwerk 180
New Cross 81
New Mexico 203, 205
New Zealand 19, 34, 206
Newchurch 48, 55
Newington 38
Nieder Mendig 115
Niedersachswerfen 129
Nieuwe Biezenweg 120

Nieuwhinkeleroord Polder 120
Nijreesbos 167
Nordhausen 5, 127–9
Nordholz 180
Norfolk 61, 69
Normandy 94, 98
North Sea 132, 133, 182, 190, 191, 195
North Weald 15, 68, 69, 158
Northfleet 193
Nottingham 207
Nucourt 96

Oberbettingen 93
Ohio 203
Orpington 84
Ossendrecht 120, 122
Ostend 106
Ostermarie 14
Oud-Beijerland 175, 176
Oxshott Woods 161
Oyle 125

Paglesham 71, 73, 80, 136
Pall Mall 19
Paris 96, 102, 154
Pas de Calais 6
Pearl Harbor 155
Peenemünde 11, 13, 172, 190
Pernis 173
Petersfield 22
Picatinny 102
Piddlehinton 8, 9
Piershil 173
Pledge Farm 58
Plomion 90, 91
Poland 67
Poole 221
Poplar 139, 193
Port-en-Bessin 87
Portsmouth 8
Preston 165
Prüm 186
Puttershoek 174, 176

Rackheath 69
Reads Island 195
Realcamp 88

Rennes 88
Richmond 140
Richmond Park 29
Rijsterbos 84
Ringweg 120
Rippsdorf 170
River Blackwater 84
River Crouch 190
River Humber 195
River Scheldt 132
River Stour 84, 187
River Thames 21, 160, 187, 193, 194
Robertsbridge 35
Rockland St Mary 69
Rohr 115
Ronne 14
Rossbach 7
Rotterdam 173, 181
Rouen 89, 90, 108
Royal Aircraft Establishment,
    Farnborough 19, 23, 34, 37, 67, 69, 79, 102, 104, 105
Rufters Wood 15, 16
Rye Hill 68

Salehurst 10
Sandhurst 48
Sarnaki 67
Saxmundham 59, 64
Schaffen 89
Schalkenmehren 167
Schelden 96
Schiermonnikoog 186
Schleswig-Holstein 111, 133
Scrums Farm 48
Seething 69
Sevenoaks 22, 197
Seymour 189
Sheepwash Farm 71
Sheffield 221
Sherrington Park 191
Shipton-on-Stour 27
Shoeburyness 143, 164
Sint-Denijs-Westrem 106
Siracourt 109
Småland 67
Snargate 38

South Kensington 19
South Shields 140
Southborough 51, 53, 221
Southend 83
Southgate 81, 83
Spa 102
Spich 176
Spijkenisse 177, 181, 182
St Johns Park 81
St Michael 49
St Paul's Cathedral 142
St Trond 98
Staffordshire 47
Stalham 61
Stampere Farm 14
Stamperegaarden 14
Stannetts Farm 71, 73
Staplecross 19, 22
Statenkwartier 71, 79
Station XII (nr Stevenage) 46
Steenbergen 120
Stepney 138
Stockholm 13, 67
Stoke Newington 139
Stone 47
Strawberry Hill Farm 19, 21, 24, 25, 27–32, 34, 35, 38, 48, 96, 220
Stream Farm 56
Suffolk 59–62, 83, 136, 193
Sunbury 140
Surlingham 69
Surrey 55, 68, 161
Sussex 10, 15, 19, 21, 24, 34, 42, 45, 55, 56, 199
Swallow's Cross 191
Swancombe 193
Swansea 142
Sweden 13, 14,
Swynnerton 47

Teddington 157
Tellis Coppice 42
Termonde 110
Teuge 186
The Firs Farm 57
Thielt 90
Thornham Magna 193

Thornham Parva 193
Thornwood Common 68
Thorpness 62, 64, 65
Three Oaks 199
Tollesbury 84
Tonbridge 56
Tondorf 115, 169, 170
Tottenham 139
Trier 115
Troisdorf 176
Tunbridge Wells 17, 18, 21, 48, 51, 56, 71, 194
Twente 118

USA 24, 39, 102, 114, 118, 127, 189, 203, 205
Usedom 190

Vanbrugh Park 81
Voorst 182

Wallasea Island 189, 190
Walsall 146
Waltham Cross 139
Waltham Holy Cross 42, 68
Washington 24, 135
Wellhead Wood 21
Welling 43–5, 47
West Ham 139
West Malling 15
Westerham 11
Westerlangstedt 133
Westerwald 186
Westminster 20
Westre 168
White Sands Proving Ground 205
Whitehouse Farm 61
Whitton 191
Wimbledon 140
Winterton 195
Wood Green 63
Woodbridge 61
Woolwich 17, 19, 23, 24, 49, 81, 90
Wrabness 31
Wright Field 203
Wrington 207
Wrotham 198

Yalding 28
Yeadon 187
York 187

Zilsdorf 186
Zwolle 95

Bomb Disposal Units
British Army
2 BD Coy 34, 124
4 BD Coy 31
5 BD Coy 19
9 BD Coy 39
10 BD Coy 60, 62
20 BD Coy 34, 52, 221
22 BD Coy 72
23 BD Coy 88, 130
24 BD Coy 88, 118, 119
25 BD Coy 90
53 BD Plat 89
59 BD Plat 90
181 BD Sec 194
196 BD Sec 34
218 BD Sec 141
224 BD Sec 88
725 BD Sec 161
CRE 2 BD Group 34
RAOC EADCU 170, 172
101 Engr Regt (EOD) 84, 85

Royal Air Force
5130 BD Sqdn 115
5131 BD Sqdn 187
5132 BD Sqdn 115
5138 BD Sqdn 118

5139 BD Sqdn 103, 111
6201 BD Flt 115, 118, 170
6203 BD Flt 128
6205 BD Flt 87, 114, 169
6206 BD Flt 104, 106, 110, 111, 114
6208 BD Flt 115, 168
6210 BD Flt 15, 104, 106, 107, 111, 114
6212 BD Flt 128
6213 BD Flt 115
6220 BD Flt 158
6224 BD Flt 168
6225 BD Flt 87, 111, 112
6226 BD Flt 192
6228 BD Flt 117, 118, 125
6229 BD Flt 124, 126
6234 BD Flt 169
6235 BD Flt 187

Royal Navy
NP 1572 131
Southern Diving Unit 2 84

US Forces
10th BD Sqdn 98
73rd BD Sqdn 99, 100
74th BD Sqdn 101
75th BD Sqdn 116
77th BD Sqdn 98
81st BD Sqdn 96

Dutch Forces
1 Coy 120, 121
3 Sect 120
Duik-en Demonteergroep Koninklijke
    Marine 182